TIME IN THE MEDIEVAL WORLD

By exploring some of the more important senses of time that were in circulation in the medieval world, scholars from a wide range of disciplines trace competing definitions and modes of temporality in the Middle Ages, exploring their influence upon life and culture. The issues explored include anachronism as a feature in earlier senses of time, perceptions of death and of the Last Judgement, time in literary narratives, constructions of time as used in the professions, and original work on the particular systems and technologies which were used for the keeping of time, such as clocks and calendars.

Chris Humphrey was British Academy Postdoctoral Fellow at the University of York; W. M. Ormrod is Professor of History at the University of York.

YORK MEDIEVAL PRESS

York Medieval Press is published by the University of York's Centre for Medieval Studies in association with Boydell & Brewer Limited. Our objective is the promotion of innovative scholarship and fresh criticism on medieval culture. We have a special commitment to interdisciplinary study, in line with the Centre's belief that the future of Medieval Studies lies in those areas in which its major constituent disciplines at once inform and challenge each other.

All enquiries of an editorial kind, including suggestions for monographs and essay collections, should be addressed to: The Director, University of York, Centre for Medieval Studies, The King's Manor, York, YO1 7EP (E-mail: lah1@york.ac.uk).

Publications of York Medieval Press are listed at the back of this volume.

TIME IN THE MEDIEVAL WORLD

Edited by
Chris Humphrey *and* W. M. Ormrod

THE UNIVERSITY *of York*

YORK MEDIEVAL PRESS

First published 2001

A York Medieval Press publication
in association with The Boydell Press
an imprint of Boydell & Brewer Ltd
PO Box 9 Woodbridge Suffolk IP12 3DF UK
and of Boydell & Brewer Inc.
PO Box 41026 Rochester NY 14604–4126 USA
website: http://www.boydell.co.uk
and with the
Centre for Medieval Studies, University of York

ISBN 1 903153 08 5

A catalogue record for this book is available
from the British Library

Library of Congress Cataloging-in-Publication Data
Time in the medieval world/edited by Chris Humphrey and W. M. Ormrod.
p. cm.
Includes bibliographical references and index.
ISBN 1-903153-08-5 (alk. paper)
1. Civilization, Medieval. 2. Time measurements – History.
3. Calendars – Europe – History. 4. Europe – Social life and customs.
I. Humphrey, Chris, 1970– II. Ormrod, W. M., 1957–
CB353.T56 2002
304.2′3′0940902–dc21 2001026924

Typeset by Joshua Associates Ltd, Oxford

CONTENTS

LIST OF ILLUSTRATIONS

LIST OF CONTRIBUTORS

Paul Brand	All Souls College, Oxford
Peter Burke	Emmanuel College, Cambridge
Mary Carruthers	New York University
Deborah Mauskopf Deliyannis	Western Michigan University
Chris Humphrey	University of York
Robert Markus	University of Nottingham
W. M. Ormrod	University of York
Ad Putter	University of Bristol
Howard Williams	Trinity College, Carmarthen

LIST OF ABBREVIATIONS

BL	London, British Library
CC	Corpus Christianorum
EETS	Early English Text Society
Le Goff, *Time, Work and Culture*	J. Le Goff, *Time, Work, and Culture in the Middle Ages*, trans. A. Goldhammer (London, 1980)
MGH	Monumenta Germaniae Historica
PL	*Patrologia Latina*
PRO	London, Public Record Office
Riverside Chaucer	*The Riverside Chaucer*, ed. L. D. Benson, 3rd edn (Oxford, 1987)
RP	*Rotuli Parliamentorum*, 6 vols. (London, 1783)
SR	*Statutes of the Realm*, ed. A. Luders *et al.*, 11 vols. (London, 1810–28)

Introduction

Chris Humphrey

The impending approach of the third millennium AD served to focus both popular and scholarly curiosity on the nature and history of time. The Middle Ages occupy an ambiguous place in this history: on the one hand recognizably pre-modern, and yet also arguably the birthplace of a modern temporal sensibility. In fact, the interest generated by the new millennium has led to a wider appreciation of the medieval origins of many of our temporal ideas, systems and technologies. For instance, the thorny issue of when one should celebrate the 'proper' beginning of the third millennium gave the sixth-century monk Dionysius Exiguus, or 'Dennis the Small' as he has been unkindly translated, a brief period of fame.[1] In this climate, works such as Dohrn-van Rossum's magisterial *History of the Hour*, Lacey and Danziger's *The Year 1000*, Gould's *Questioning the Millennium* and (for a later period) Sobel's *Longitude* have brought apparently 'specialist' information about time to a wider audience – and even to the best-seller lists.[2] Appropriately enough for a subject area that is dominated by the imagery of circularity, our present position in time has led to a renewed interest in how and why our experience of temporality should be like this at all.

The papers collected in the present volume represent one way of answering this perennial question. With the exception of the contribution from Dr Ad Putter, the papers collected here were given as lectures in the York Medieval Seminar series for 1999. For this series, both established and younger scholars presented findings from their research on the use and representation of time in the medieval world. Rather than trying to define what medieval time is or was, the papers in the series and consequently in this volume set out to explore what medieval people *did* with time: the emphasis is on the practical utility of temporal schemes as opposed to their

[1] G. Declercq, *Anno Domini: The Origins of the Christian Era* (Turnhout, 2000), p. 97. Rather than reflecting a diminutive stature, Dionysius's epithet was a humble self-reflection on his intellectual stature.

[2] G. Dohrn-van Rossum, *History of the Hour: Clocks and Modern Temporal Orders*, trans. T. Dunlap (London, 1996); R. Lacey and D. Danziger, *The Year 1000: What Life was Like at the Turn of the First Millennium* (London, 1999); S. J. Gould, *Questioning the Millennium: A Rationalist's Guide to a Precisely Arbitrary Countdown* (London, 1997); D. Sobel, *Longitude: The True Story of a Lone Genius who Solved the Greatest Scientific Problem of his Time* (New York, 1995).

absolute characteristics, whether in literary narratives, legal definitions or religious beliefs. The chronological scope of the papers, from the early Middle Ages to the early modern, suggests that it is futile to try to define 'medieval time' in absolute terms: then, as now, time was a malleable and relative medium.

One paper in the volume amply demonstrates this point for seventh-century England. Howard Williams discusses current anthropological and sociological approaches to death and social memory, and considers how these perspectives have been used in the interpretation of ritual and funerary practices in past societies. For instance, one perspective on burial sites is that they etch an indelible memory of the interred person or persons on the landscape. Using a case study of the high-status burial barrows at Sutton Hoo in Suffolk, Williams argues for a more sophisticated interpretation of the role of burials in creating perceptions of time and memory. For instance, the continuing re-use of graves at Sutton Hoo demonstrates how interments captured more than just one particular point in time for posterity: in fact the way that an interment was located in time changed according to its role in current social practice. We are used to thinking about how a society's graves reflect its view of eternity and the afterlife; now we are also urged to reflect upon the afterlife of graves themselves.

Robert Markus's study of the end of time in early Christian thought continues the theme of how apparently fixed points in time can be transformed by the passage of time. Early Christians were convinced that the end was nigh, but they did not know when: for subsequent generations the embarrassing absence of apocalypse meant being ready for the end at any time. It is this sense of 'imminence with indefinite postponement' that Markus sees as defining the theology of later thinkers like Pope Gregory, and which lies at the heart of the Christian understanding of time and its ending.

Articulating one's belief in the fundamental ambiguity of temporality may be a powerful statement of faith, but it is unlikely to hold up as an excuse for avoiding your annual tax bill. In an illuminating exploration of year-dating conventions in the early Middle Ages, Deborah Mauskopf Deliyannis considers some of the more prosaic forms of time keeping in use in the sources. Ecclesiastics, notaries and historians had a wealth of dating conventions to choose from, including regnal, consular and indiction years, as well as counting from fixed points such as the birth of Abraham or the founding of Rome. Given the absence of a common standard to refer to until the AD system became widespread, Deliyannis defends the record of early historians when it came to getting dates wrong: 'perhaps we should rather marvel that they got any dates right'.

Two papers in this volume address the representation of time in literary texts of the later Middle Ages, offering new insights into the use of time in medieval narrative. Mary Carruthers considers the way that authors like

Chaucer and the *Gawain*-poet used the tense known as the 'historical present', narrating events ostensibly in the past in the present tense. Not only did medieval authors display a seeming disregard for whether they used the past or present tenses when narrating historical events, they also went so far as to put their narrators into the story by describing things that they saw or heard. This apparent indifference to proper temporal sequence has been seen by some as embodying a characteristically medieval attitude towards time, lacking the clear sense of anachronism that defines the modern age. Carruthers offers a more sophisticated view of this situation by suggesting that while the past is the time of narrative, the present is actually the tense of narrative: it is only modern literary convention that prefers past events narrated in the past tense. Techniques such as tense switching and a first-person perspective were therefore ways in which medieval narrators could convincingly remember and guarantee the stories that they told.

Ad Putter also finds the modern critic all too ready to accuse the medieval author of a lack of respect for temporal continuity in narrated time. Exploring a particular textual 'problem' in each of two medieval romances, Putter shows how a fuller understanding of the medieval calendar enables convincing solutions to be proposed in both cases. The supposed medieval indifference to time rests rather on modern misreading and scribal error, and with these corrected it is our modern indifference to medieval time that is most starkly revealed.

In his analysis of lawyers' time in the later Middle Ages, Paul Brand presents an extensive review of how the profession used and viewed time in this period. The two principal themes of the study are present time, taking in the legal terms and return days, and past time, the way that lawyers ascertained when the matters in question had taken place. The wealth of detail presented here will be welcomed by the legal specialist and appreciated by anyone wishing to know more about this fascinating dimension of medieval law. The discussion extends to those unfortunate litigants whose cases were lost because they had failed to specify the exact saint's day on which the alleged crimes took place (St Stephen and St Michael the Archangel had three feast days each, for example).

Chris Humphrey addresses an apparent paradox in the evidence for the adoption of clock-based ways of reckoning time in late medieval England. While the European evidence and English literary sources attest to a strong connection between towns and clocks, a closer examination of the city of York suggests that its citizens lagged behind their ecclesiastical neighbours when it came to installing these expensive timepieces. Such a paradox disappears when the fuller picture is examined: what is stake is not the simple acquisition of clocks, but rather their use in structuring the daily life of a community that can hear and make use of its signals. It was in this latter area that the citizens of medieval towns led the way, since they used the potential inherent in clock technology to create a distinctive rhythm for their lives, a rhythm that has endured to the present day.

The volume concludes with Peter Burke's reappraisal of the new sense of the past expressed by artists and scholars in fifteenth- and sixteenth-century Europe. Burke considers a range of ways in which medieval people expressed a sense of continuity with the past, including the 'Roman soldier test', where the soldiers at the tomb of Christ are represented in contemporary dress. Such a perspective on the past differs from the attitude of the later artists who valued the accuracy of historical detail, and from the view of the scholars who looked back to the 'Middle Ages' that we still refer to today.

In drawing some common threads from the broad range of ideas and observations collected in this volume, one cannot but help be struck by the recurring themes of identity and community. We are all used to the idea that time is relative; what the papers in this volume have further demonstrated is that time can be a way of building social relationships. Ask 'What is the time?', and one creates a shared reference point in the otherwise subjective flow of individual experience. Ask 'What is time?', and one has a way of exploring something of what binds communities of faith, status and common purpose together, both in the past and also for the future.

1

Year-Dates in the Early Middle Ages

Deborah Mauskopf Deliyannis

The method of naming years known as Anno Domini (AD) dating has been in use since the sixth century. Its basic premise is that Jesus was born in the year 1; all subsequent years, numbered consecutively, are years of the Incarnation of Christ. A consecutive, numerical system such as this one allows us to calculate automatically how much time has elapsed between two dated events, and how they are related in time to other events. This may seem obvious: as modern historians, we are accustomed to working with AD dates. If we encounter a date in some other system, we instinctively reach for a table of correspondences so that we can assign the correct AD date, before proceeding further.

Even though AD dating seems so handy to us, it took several centuries after its introduction before it became the standard dating system in Europe. Before this, a number of different systems for naming years were in use, most of which were not continuously numerical. Without a numerical set of year-dates, it is not at all easy to calculate the numerical passage of time, although it is certainly possible to do it. What I think this means is that when they were used, year-dates were not intended to situate events on a continuous time-line, but rather served other purposes. Time itself, especially past time, was not measured in years, but in other units. It could be counted, but only for very specific purposes.

There were several groups of people in society who needed to know when in time they were. The clergy relied upon calendars to tell them when holidays, particularly Easter, were to be celebrated. Scribes and notaries who drew up legal documents needed to be able to put dates on the documents. And (by far the smallest group) people writing history needed to be able to situate events according to a chronological framework.[1] In this essay, I would like to present the evidence for how each of these groups used year-dates, with an emphasis on the historians: to discuss the ways that the

[1] A fourth group of people who used dates were authors and/or carvers of inscriptions; I have not examined this evidence, but an extensive discussion of the dates used on inscriptions in Britain, Gaul, and Spain between 350 and 750 can be found in M. Handley, 'The Early Medieval Inscriptions of Britain, Gaul and Spain: Studies in Function and Culture' (unpublished Ph.D. dissertation, University of Cambridge, 1998), esp. ch. 6.

uses of dates in society influenced how a historian could write, and to present some conclusions about how the past was organized.

DATING SYSTEMS

We should first review briefly the forms of dating available in the early Middle Ages. In the Roman Empire, there were many different ways of naming years, some used officially by the government, others purely local, and many of these systems continued to be used throughout the Middle Ages.[2] The three most common were the following.

Consular years, in which the year was named after the two consuls of Rome, had been used since 450 BC, and continued through the imperial period.[3] The last named consul in Constantinople, Basilius, held office in AD 541; for the next twenty-six years, the years were numbered as years *post consulatum Basilii*, 'after the consulship of Basilius', with a number. In 567, the consulship in Byzantium became an office of the emperor, and after the first year of his reign, the *post consulatum* designation was used until the accession of the next emperor.[4]

Regnal years need no further explanation, except to note that after the end of the Roman empire in the west, regnal years might be of a king, a duke, a bishop, or another form of authority.[5]

Finally there were indiction years.[6] This is a system whereby each block of fifteen years, starting in AD 312, makes up an indiction cycle. A year is identified as part of the current indiction (for example, 'in the tenth year of the indiction'), but the cycles themselves are not numbered.

None of these dating systems is numerically continuous, and of the three systems, only indiction years are not dependent upon a specific political regime. Continuous numerical systems did exist, although they were less

[2] For a comprehensive summary, see R. D. Ware, 'Medieval Chronology: Theory and Practice', in *Medieval Studies: An Introduction*, ed. J. M. Powell, 2nd edn (Syracuse, NY, 1992), pp. 252–77.

[3] In the fourth century, one consul was named for Constantinople and one for Rome; after the fall of the empire in the West, the Roman consul was named by the Ostrogothic kings, until 534. See H. Fichtenau, '"Politische" Datierungen des frühen Mittelalters', in *Intitulatio*, ed. H. Wolfram, 2 vols. (Vienna, 1973), pp. 453–548 (pp. 459–63).

[4] Curiously, in most parts of western Europe consular years ceased to be used after 567, but in the kingdom of Burgundy, the *post consulatum* formula was used on inscriptions until as late as the mid-seventh century: see Handley, 'Early Medieval Inscriptions', pp. 163–87.

[5] It might also be noted that when there was more than one emperor (such as a junior emperor), there would be more than one designation. For example, 'the sixth year of Leo and the fourth year of Constantine' (AD 722).

[6] Justinian's *Corpus Iuris Civilis* called for dating by the indiction (*Novellae* 47.i: see Ware, 'Medieval Chronology', p. 264), and indiction years were used in western Europe until the fourteenth century.

commonly used: indeed, they tended to be used in historiography rather than in daily use. There was the numerical reckoning *ab urbe condita*, from the founding of Rome. The Greek Olympiads began in 776 BC and numbered each group of four years. In Spain, starting in the third century, years were numbered according to an era that we call the Spanish Era, which begins in 38 BC.[7] In the second and third centuries AD, both the Jews and the Christians started to count years from the Creation according to the Bible, which is known as the *annus mundi*. The *annus mundi*, while useful for theological speculation about the Ages of the World, was subject to varied interpretation, and several scholars recalculated the base date of the Creation.[8] Recognizing this, the historian Eusebius based his chronology on an *annus Abrahamae*, finding it easier to count from the birth of Abraham than from the Creation.

In most contexts in which dates are used, only the present or the past needs to be named. In a situation in which the future must be described, a consecutive numerical system must be used. In the Christian world, such a situation arose in connection with the controversy over the dating of Easter, and led ultimately to the creation of the *annus incarnatione domini*, or AD year.

ECCLESIASTICAL USE

As is well known, the calculation of the date of Easter was controversial in the early Middle Ages, and those who claimed to be doing it the 'right' way needed to have a way of disseminating the information so that the feast could be observed correctly. The solution that was developed was to have the dates of Easter calculated for some period of time into the future, written on a table, and sent out to the churches of Europe. In order to create such a table, a numerical system of years was required; in order to use the table, one would need to know the year in which one was living.

While indiction cycles were too short to be used for long tables, regnal or consular naming systems could obviously not be used for the future, since tenure was unknown. In Alexandria, where such tables were first constructed, years were dated from the beginning of the reign of the Emperor Diocletian, corresponding to AD 284. Naming years after Diocletian, a great persecutor of Christians, was not popular, and the first table to be widely diffused in the west, compiled by Victorius of Aquitaine and giving the dates of Easter for the first 532 years after the Crucifixion, was dated in Years of the Passion and in consular years. The *annus passionis*, beginning in AD 29, was

[7] See Handley, 'Early Medieval Inscriptions', pp. 188–201.

[8] See esp. R. Landes, 'Lest the Millennium be Fulfilled: Apocalyptic Expectations and the Pattern of Western Chronography 100–800 CE', in *The Use and Abuse of Eschatology in the Middle Ages*, ed. W. Verbeke, D. Verhelst and A. Welkenhuysen (Leuven, 1988), pp. 137–211, for a discussion of the various different *anni mundi*, who used which one, and when.

certainly appropriate for use on Easter tables, but when Dionysius Exiguus constructed a new set of tables, which were to become the standard in Europe, he instead used AD and indiction years.[9]

Every major church must have had an Easter table, and thus in all parts of Europe AD dates, correlated with the indictions, must have been available. In many places these tables were gradually annotated with noteworthy events for each year, ultimately producing Easter annals. This is not the place to go into the history of the annal,[10] but it is worth considering the idea that this annotation was done to link the AD and indiction dates – which were universal but not well understood – with local events as an *aide memoire*, so that future clergy would know what year they were in. For that matter, it might even have been a way of ticking off each year as it passed.

One final thing to note about clerical use of dates is that, despite the availability of Easter tables, it was considered desirable for individuals to be able to calculate the date of Easter themselves, based on the lunar and solar calendars. A type of text, the *computus*, was devised, first by Cassiodorus, that explained how both sorts of calendars worked, and how to make the necessary conversions and calculations between them.[11] In these, two types of years are mentioned: the indiction and the Incarnation, and usually a mathematical way of converting between the two is given. This implies that the people using a *computus* would be familiar with either indiction or AD years, the two dating systems used on Dionysian Easter tables. Interestingly, Bede, in his *computus*, specifically says that indiction years are used to avoid complications arising from regnal years: when a king dies, one *historicus* might name the year after him, and another after his successor, whereas indictions are regular.[12] Note that Bede does not claim this universal application for AD years.

[9] See *Bedae Opera de temporibus*, ed. C. W. Jones (Cambridge, MA, 1943), pp. 116–22, for a full discussion of this question. Evidence that AP tables were still available in the eighth century is discussed by M. Miller, 'Consular Years in the *Historia Brittonum*', *Bulletin of the Board of Celtic Studies* 29 (1980), 17–34. It is not clear why AD years should have been preferred to AP years, but one possibility is that there were at least two different ways of calculating the AP, one used by Eusebius-Jerome that dated the Crucifixion to Tiberius's eighteenth regnal year, and one used by Prosper of Aquitaine using Tiberius's fifteenth year. See M. Humphries, 'Chronicle and Chronology: Prosper of Aquitaine, his Methods and the Development of Early Medieval Chronography', *Early Medieval Europe* 5 (1996), 155–75 (pp. 159–60).

[10] See especially M. McCormick, *Les Annales du haut Moyen Age*, Typologie des Sources de Moyen Age Occidental 14 (Turnhout, 1975), and *Bedae Opera*, ed. Jones, esp. ch. 7.

[11] See A. Borst, *The Ordering of Time: From the Ancient Computus to the Modern Computer*, trans. Andrew Winnard (Chicago, 1993). R. McKitterick, 'Constructing the Past in the Early Middle Ages: The Case of the Royal Frankish Annals', *Transactions of the Royal Historical Society*, 6th s. 7 (1997), 101–29 (p. 108) notes that Charlemagne's *Admonitio generalis* of 789 required the clergy to know the *computus*.

[12] *Bedae Opera*, ed. Jones, p. 268.

NOTARIAL USE

The second group of people who used dates regularly were scribes and notaries. The *datatio* was a standard part of most legal documents, so the authors would have to know how to calculate a date. Most of the documents from the early Middle Ages survive not as originals, but as later copies, and as they exist now, a majority of the documents lack dates altogether. Perhaps when the items were recopied, the dates were omitted, especially if the later copyists could not understand them. Or perhaps such items simply were not given year dates, which is interesting also. Despite these problems, it is clear that dating patterns do exist, and students of diplomatic have recognized which formulae were used from kingdom to kingdom and from period to period.[13] We will briefly consider these for Italy, Francia and England.[14]

Italy

For the purposes of dating of documents, there are two main regions in Italy: the territory that was imperial and became Byzantine, and the territory that became Lombard. Dating of documents in Byzantine areas reflected imperial practice; the way this changed over time can be seen perhaps most clearly in papal archival documents. Throughout the fifth and into the middle of the sixth century, consular dates are used, and then, following Justinian's standardization of dating practices in his *novella* 47 of 537, the imperial year, consular year, and indiction. The formula *post consulatum Basilii* was used from 541 until 567, when the consulship was made an office of the emperor; until the late eighth century, the date is given in imperial regnal years combined with the indiction. Then, during the iconoclastic controversy, some documents are dated only with the indiction, deliberately omitting imperial regnal years. In the 770s, the imperial years were dropped in favor of papal regnal years plus the indiction, and then Frankish regnal years plus the

[13] A comprehensive treatment of the dating of documents from Germany and Italy is found in H. Bresslau, *Handbuch der Urkundenlehre für Deutschland und Italien*, 3rd edn, 2 vols. (Berlin, 1958), II, 393–478, esp. pp. 406–41. A full discussion of the political meaning of dating *formulae* from Late Antiquity to the late Carolingian period, with a focus on the Carolingian territories, is given by H. Fichtenau, '"Politische" Datierungen'), pp. 453–548.

[14] The evidence from Visigothic Spain is discussed by Handley, 'Early Medieval Inscriptions', pp. 190–201. He notes that Church councils were dated by the Spanish Era, as were, apparently, some legal documents mentioned in the *Formulae Visigothicae*, but the laws were dated by regnal years. In southern Spain, many epitaphs were dated by the Era, and Handley argues that this was a specifically Catholic, un-Arian, act.

indiction.[15] Not until the mid-tenth century do AD years start being used in papal documents.[16]

There are also numerous surviving documents from Ravenna. From the fifth to the seventh centuries, the dating formulae are identical to those used in Rome. In the eighth and ninth centuries, the formula is still regnal plus indiction year, where the regnal year could be of Lombard kings, Carolingian kings/emperors, or popes. Becoming common in the 870s, and standard in the 920s, was the combination of papal regnal, imperial (non-Byzantine) regnal, and indiction years. This formula was continued through the tenth century.[17]

In the Lombard kingdom, documents were also consistently dated by royal regnal year plus indiction, obviously in imitation of the Byzantine form. A similar formula was used in the dukedom of Spoleto, where the date usually contained the regnal year of the duke, and also, interestingly, the name of the gastald, as well as the indiction. Sometimes the regnal year of the king, Lombard or later Carolingian, would also appear.[18]

AD dates on documents appear occasionally in Italy even in the Carolingian period, and start becoming more common in the early tenth century. There is an upsurge in their use after the Ottonian conquest, especially in papal and imperial documents. However, even after 962, most documents not directly connected with the imperial court are still dated by regnal years plus the indiction.[19] It is only around 985, shortly after the death of Otto II, that documents all over Italy start being dated by AD years.

Francia and the Carolingian empire

In the Merovingian kingdoms, charters and documents are dated by the regnal year of the king only. The only place in which some indiction dates appear are acts of Church councils from the sixth century, alongside consular or regnal dates. This use of indiction years probably reflects the ecclesiastical influence of Rome.

[15] See Fichtenau, '"Politische" Datierungen', pp. 487–500, for a detailed analysis of the complicated changes in dating formulae in this period.

[16] See R. L. Poole, *Studies in Chronology and History* (Oxford, 1934), pp. 172–84, who notes that Byzantine imperial regnal dates were used on papal documents until 772; for the next nine years there are no dated documents extant, but in 781 one appears dated to the papal regnal date and the indiction. The first Carolingian regnal date in papal documents occurs in 798; imperial dates were then used until 905, and from 962 to 1047.

[17] There is an interesting period from 984 to 993 in which the imperial dates are omitted.

[18] For the fluctuations in dating formulae in the Spoletan dukedom, see D. Bullough, 'The Writing-Office of the Dukes of Spoleto in the Eighth Century', in *The Study of Medieval Records. Essays in Honor of Kathleen Major*, ed. D. A. Bullough and R. A. Storey (Oxford, 1971), pp. 1–21.

[19] The only documents in the Ravenna archive which are dated using AD are imperial charters; all the ordinary documents are dated the usual way.

It is often noted that AD dates first make their appearance in Francia at the beginning of the Carolingian period. Actually, the first *capitulae* listed in the MGH *Leges* that are dated with AD years come from the days when Pepin was the mayor of the palace; presumably the alternative, Merovingian regnal dates, somehow did not fit in a charter issued by the mayor.[20] Nevertheless, AD dates are still extremely rare. For the most part, Pepin and then Charlemagne continued the Merovingian practice of dating by regnal years alone.[21] After 800, the years are given as regnal years of Charlemagne's imperium, sometimes also with his years as the king of the Franks and the Lombards. Indiction years appear, and become standard, after 802, presumably on the Byzantine imperial model, and AD years appear occasionally, more frequently in the 840s and 850s.[22] It was in 876 that Charles the Fat and Louis the Younger, after the death of their father Louis the German and perhaps in opposition to their older brother Carloman's imperialist designs, officially began to use AD years in their official documents, and the practice spread through East Francia. In West Francia, AD dates were only introduced by the first non-Carolingian, Odo, in 888.[23]

England

Finally, in Anglo-Saxon England, a completely different pattern appears. By far the most common formula for dating documents is AD plus the indiction. In Latin documents, regnal dates are somewhat more common than in Anglo-Saxon documents, although almost always used together with AD and indiction. It has been traditional to credit Bede with introducing AD dating to England. However, AD by itself, or AD plus indiction, appears on even the very earliest documents, from the seventh century.[24] Kenneth Harrison has published a detailed study of this issue, and concludes that dating with AD was indeed in use in the seventh century, that is, before Bede wrote his chronological works.[25]

[20] See, for example, M. Innes and R. McKitterick, 'The Writing of History', in *Carolingian Culture: Emulation and Innovation*, ed. R. McKitterick (Cambridge, 1994), pp. 193–220 (p. 198). As F. Lifshitz, *The Norman Conquest of Pious Neustria: Historiographic Discourse and Saintly Relics, 684–1090* (Toronto, 1995), p. 58, notes, it gave the Carolingian usurpers authority not tied to the Merovingian dynasty.

[21] McKitterick, 'Constructing the Past', p. 104, n. 16, notes that the *diplomata* of Charlemagne of 783, 788 and 791 with AD dating formulae survive in later copies and the AD formulae are thought to be later interpolations.

[22] See Fichtenau, '"Politische" Datierungen', pp. 536–8.

[23] See Fichtenau, '"Politische" Datierungen', pp. 537–9.

[24] In most cases, even on Anglo-Saxon documents, the dating formula is written in Latin with the word *indictio*. However, on two Anglo-Saxon documents different words are used: in one case *tacencircole* and in the other *gebongere*: *Anglo-Saxon Charters*, ed. A. J. Robertson (Cambridge, 1939), p. 864 (no. 11) and p. 904 (no. 18).

[25] K. Harrison, *The Framework of Anglo-Saxon History to A.D. 900* (Cambridge 1976).

Summary

Certainly every notary would have had a list of dates in the system or systems in use. In the Roman Empire, lists of the consuls were a standard notarial tool. Lists of the lengths of reigns of rulers also survive from the early Middle Ages: for example, they are found at the end of some of the Germanic lawcodes.[26] Presumably at any given moment, a notary would know who the king was, and how many years he had reigned. Or, in some cases a notary did not know who the king was, and had to use a *post regnum* date referring to the previous king. In Italy, these lists must have been correlated with the indiction years, in continuation or imitation of Byzantine practice. In the Frankish kingdoms, however, all that was required was the list of kings. In spite of the availability of AD and indiction year-lists from the Easter tables, only in England was the combination of AD plus indiction years used on documents. If, as has been suggested by Susan Kelly, legal documents in England were first used in ecclesiastical contexts, and then spread to the secular population, the use of Easter-table dates is somewhat understandable,[27] although it should be remembered that AD dates were not used by the Church elsewhere (as in Rome), so in fact this was an innovation unique to England. Given the number and fluid nature of Anglo-Saxon kingdoms, dating by some system other than regnal years perhaps allowed the date to be more widely recognized throughout England.

Year-dates imply a sense of the passage of time. If a document is dated, we tend to assume that it is to let you know when it was written, so that you might know (in some future) how long ago a particular transaction took place. The phrase 'how long ago' implies a certain understanding of the past. However, without a numerical dating system like AD years, it is difficult, although not impossible, to determine specifically how long ago something was done.

On the other hand, a regnal date is also a statement of authority, indicating who the ruler is. As Fichtenau notes, because dating systems were so closely tied to political regimes, the use of a date on a document serves to validate the ruler's authority in a very particular way, by naming time after him.[28] That this was an important function of the date can be seen in the situations in which a dating system was changed, parallel to a change in regime, as, for example, with the end of Byzantine rule in Italy. The adoption of indiction dating by

[26] Found at the end of the *Lex Visigothorum* and the *Lex Salica*: see MGH *Leges* Sectio I, 1.

[27] S. Kelly, 'Anglo-Saxon Lay Society and the Written Word', in *The Uses of Literacy in Early Medieval Europe*, ed. R. McKitterick (Cambridge, 1990), pp. 36–62 (pp. 42–3).

[28] See Fichtenau, '"Politische' Datierungen'. Handley, 'Early Medieval Inscriptions', p. 161, notes an example of a regnal date on a Spanish inscription in which the reign is begun at the year the king went into rebellion against the former king, not the year that his rulership was recognized; thus, 'the use of regnal dates on these inscriptions acted to legitimize the years spent by Athanagild as a usurper.'

Charlemagne after he was crowned emperor demonstrates an awareness of the association of indiction dates with imperial regnal years. In many cases, dates on documents might have been less to place the item in long-term time than to indicate who was in charge. This is a very short-term purpose.

By contrast, naming the year after the Incarnation of the Lord gives the date a universal quality, and also a sacred one. It is tempting to identify the adoption of AD dating on documents with a desire to sacralize time, to link it to a programme of *renovatio* such as Charlemagne's. However, the evidence suggests instead that AD dates were adopted in moments of political fragmentation: in seventh-century England, in late ninth century Francia, in late tenth century Italy. At times in which one ruler's authority was becoming limited, or was no longer recognized, people cast about for a suitable replacement for regnal years. This can be seen in the Ravennate documents, where for the later ninth and tenth centuries the dating formulae fluctuate rapidly between papal regnal, imperial regnal, and indiction years. AD provided a default: available to everyone via Easter tables, it was politically neutral and spiritually appropriate.

HISTORIOGRAPHICAL USE

If dates on documents did not serve primarily to locate the event in time, what, then, of the dates used by historians, whose business was to situate events in time? The plethora of dates in use in the early Middle Ages made the task of an early medieval historian rather difficult. Early medieval historians, like their modern counterparts, had to do research in order to write their histories. In most cases, they relied on earlier written histories, literary texts, documents, monuments and oral testimony, each of which might contain dates that were named according to different criteria. They selected from these sources information that they felt was relevant to include in their own work, and arranged it according to their work plan. The ways that information was selected and incorporated varied from author to author, but one issue was common to all: the problem of situating events in a chronological framework. Since our understanding of the history of the early Middle Ages depends heavily upon the few historical texts that survive, it is crucial that we consider what kinds of chronological constraints the authors might be working under, before we criticize their inaccuracies.

There were two parts to a historian's task: first, he had to work out a chronological structure by which he could organize his work, using whatever dates he felt relevant; and second, he had to figure out how information from his sources fitted into this chronological scheme. The two were, of course, interrelated; depending on what sources of information an author had, he would shape his chronology accordingly.

First let us consider the chronological structure. Authors of year-by-year annals had a relatively simple task. They needed a framework of year-dates, but once they had this, they simply had to write in major events for each year. The earliest of the annals seem to have grown out of date-tables compiled for other purposes: Easter tables, as I have already mentioned above, and the consular lists used by notaries, were both annotated and gradually developed into their own historiographical genres.

Medieval authors of chronicles had their task made somewhat easier because of the chronographical base created by Eusebius, who synchronized biblical and classical history in his *Chronicon*. Chroniclers after Eusebius created further lists of years, and these texts were quite widely available. Eusebius started with the birth of Abraham, which he believed corresponded to the reign of King Ninus of Assyria, and he based his chronology on the *annus Abrahamae* whose date of the birth of Abraham corresponds to our 2016 BC. He also named each year according to the Greek Olympiads, and kept a continuous regnal list for the Roman emperors. Jerome's translation of the *Chronicon* into Latin was influential in western Europe, and is mentioned by name by many subsequent historians. Interestingly, none of his continuators kept the same chronological framework. Several who wrote in the sixth century switched over to using consular years, although they claimed to be continuing Jerome; in effect, they were now writing consular annals.[29] Others used primarily imperial regnal dates, supplemented by a universal era: *annus mundi* in the case of Bede and Isidore of Seville, Olympiads for Hydatius, and consuls plus *annus passionis* for Prosper of Aquitaine.[30] In fact, both Bede and Isidore (and others) start their chronicles not with the Assyrians, but with the Creation, and thus an *annus mundi* is more appropriate for counting forward from the beginning of the text. Even when a chronology was predetermined, putting information from other sources into it was not necessarily a simple task, especially if the year-dates in the source were different from those used by the chronicler, as Mark Humphries has recently shown for Prosper of Aquitaine.[31]

Both chronicles and annals are forms of historical writing in which the year is the main organizing feature. Narrative histories, on the other hand, are not generally organized on a year-by-year basis, but by some other principle. For these texts, depending on the author and his sources, year-dates are of considerably less importance.

Even authors who wrote chronicles as well as narrative history often did

[29] John of Biclar, Victor of Tunnuna, Marius of Avenches and Cassiodorus.

[30] John Malalas used an unusual *annus mundi* for the first part of his chronicle, then imperial regnal dates; he also included regnal dates of various places (mostly in Greece), consular years, indiction years, the Era of Antioch (based on 49 BC), a few dates from the Ascension, and various other fragments. See E. M. Jeffreys, 'Chronological Structures in Malalas' Chronicle', in *Studies in John Malalas*, ed. E. Jeffreys, B. Croke and R. Scott (Sydney, 1990), pp. 111–66.

[31] Humphries, 'Chronicle and Chronology'.

not use the same systems of dating in the different types of works. Eusebius's *Ecclesiastical History*, for example, contains only imperial regnal dates, which are certainly to be found in the chronicle, but apparently Eusebius did not feel the need to tie his narrative history to a universal chronological framework. Isidore of Seville based his chronicle on the *annus mundi* and imperial regnal years, but in his *Histories of the Kings of the Goths, Vandals and Sueves*, he uses the Spanish Era as well as imperial regnal dates. Bede's chronicle was based on his new calculation of the *annus mundi*, but in his *Historia ecclesiastica* he provides AD dates, often together with an indiction date, as well as dates *ab urbe condita* and for reigns of emperors and kings.[32]

Other authors of narrative texts use a similar plethora of dating mechanisms. Gregory of Tours uses few dates in his text. For the early part of the *Historia Francorum* in Book 1, from the Creation up to about 400, imperial regnal dates are sometimes given. In the next three books, covering the years until 576, there are only occasional references to the deaths of Frankish kings and the lengths of their reigns, but no external dates to these events are provided. However, Gregory was clearly aware of chronology, based on the world chronicle tradition; at the end of Books 1, 2 and 4 he provides a summing-up of the number of years elapsed from the Creation until that moment in his history. Starting in Book 5, the history becomes a year-by-year annal, based on the regnal years of Childebert II.[33] Gregory concludes the entire work with a grand summing-up of the length of time from the Creation until the present, defined as 'the twenty-first of my own consecration, which is the fifth year of Gregory, Pope of Rome, the twenty-third of King Guntram and the nineteenth of Childebert II', as 5,792 years. Gregory's continuator, Fredegar, used *annus mundi* as well as royal regnal dates, following Gregory's example.

In Italy, different patterns emerge. In the earliest parts of the Roman *Liber pontificalis*, written in the 530s, dates for the papal reigns are given in consular years; in the later sections, written more or less continuously from the 640s to the late 880s, the dates are given first in imperial regnal years, then in Carolingian regnal years, always with the indiction as well. Jordanes, in his *Romana*, used *annus mundi*, imperial regnal, and *ab urbe condita* dates, but in the *Getica* used only the lengths of reigns of the Gothic kings. Paul the Deacon

[32] Landes, 'Lest the Millennium', pp. 200–1, suggests that the reason that AD dates were used, and adopted by other historians throughout Europe, was that the *annus mundi* years were approaching the year 6000, which some felt would be marked by the Second Coming, although the orthodox position opposed this belief. By switching away from using the *annus mundi*, the question would be avoided.

[33] At the end of each year, Gregory provides a summing up of the weather, natural disasters and other unusual occurrences for the year. In his list of the bishops of Tours, imperial regnal years are provided for the first four bishops, but after that there are no dates, only the lengths of their reigns. Gregory even notes, 'I have not recorded the years in order as in a chronicle, for I have not been able to discover the exact intervals between the consecrations.'

provides year dates for only six events in the whole *Historia Langobardorum* (hereafter *HL*): of these, all six have indiction dates, one is given an Incarnation date (the invasion of the Lombards into Italy), and one is an imperial regnal date (the death of Phocas, taken from Bede).[34] Otherwise, time is marked, as in Gregory's work, by the deaths of kings.

There are a few other examples. The ninth-century parts of the *Historia Brittonum* used the *annus passionis* of Victorius of Aquitaine's Easter tables, with the garbled addition of some consular dates.[35] Finally, some texts, such as the *Liber historiae francorum*, Gildas's *De excidio*, and Victor of Vita's *Historia persecutionis Africanae provinciae*, contain no dates at all.

If an author starts his history with the Creation of the world, as do Gregory of Tours, Fredegar and Jordanes in the *Romana*, then it is appropriate to use *annus mundi* throughout the text, which serves to situate the later history. If, however, the text starts not with the Creation but with a specific historical event, as in all of the other examples I have mentioned, then using the *annus mundi*, or doing calculations of length of time from the Creation, is irrelevant.

In fact, most of the other historiographical texts have, at best, self-contained chronological structures: by far the most common temporal references in narrative histories are not dates, but phrases such as *hoc anno, eo tempore*, and *huius temporibus*.[36] It is very telling that in most editions and translations of narrative histories, the footnotes and margins are full of AD dates for our benefit. This indicates that universal year dates as such were of secondary importance to early medieval historians; the important thing was to understand the sequence of events in relation to an internal chronology, such as the reigns of the local kings or popes, not how much time passed between one event and another.

The other thing that can be noted about the ways these authors used dates is that the type of dates used correspond closely to the dates found in the sources presumably used by the author, including documents from the region. If an author used several sources with different types of dates, all those types are found in the text. Even assuming that each author had some sort of chronological framework for his history, how did he take information from his sources, both textual and documentary, and situate it in his framework? The question of how historians used their sources is a vast one, and one which I am only beginning to study in detail. I would like to demonstrate one case in which it can be seen that considerations of dating by

[34] Paul the Deacon, *HL* ii. 7 (invasion of Lombards into Italy), ii. 25, iv.28, iv. 29 (death of Phocas), v. 11, vi. 5.

[35] Miller, 'Consular Years', pp. 30–1, claims that Victorius's scheme would have still been taught at York in the ninth century, because it was still in use in parts of Francia.

[36] J. Favrod, 'Les sources et la chronologie de Marius d'Avenches', *Francia (Paris): Forschungen zur westeuropäischen Geschichte* 17 (1990), 1–21 (p. 21), notes that this is seen in the chronicle of Marius of Avenches, as a way of providing information from disparate sources with a chronological unity.

year are almost entirely absent, because we have both the text and many of the sources used in its composition. The text is Agnellus's *Liber pontificalis ecclesiae Ravennatis* (hereafter *LPR*).

Agnellus's historical methodology

Agnellus wrote his history of the bishops of Ravenna starting around 830 and continuing for at least seventeen years. The work, modelled on the Roman *Liber pontificalis,* consists of a Life for each bishop of Ravenna from the apostolic founder, Apolenaris, to Agnellus's own day. The identifiable sources that Agnellus used as information are the following: a list or lists of the bishops of Ravenna; the Roman *Liber pontificalis;* Paul the Deacon's *Historia Langobardorum;* a consular annal and/or something that went under the name of the *Chronicon* of Maximian; the *Passio* of St Apolenaris; a volume of the sermons of Peter Chrysologus; documents and letters of the Ravennate church; inscriptions and dedications on buildings and precious objects; and oral tradition of various sorts. Agnellus also had imitated and borrowed from various works by Gregory of Tours and Gregory the Great.

Each of these sources contained some, all, or no dates in the several systems mentioned previously. In all these works, there is only one AD date, the one given by Paul the Deacon for the arrival of the Lombards in Italy in 568. Paul otherwise gives only a few indiction and regnal dates; Agnellus's annal would have had consular dates; the *Chronicon* of Maximian, if it was indeed something like a continuation of Eusebius/Jerome, might have had *annus mundi* dates; the *Liber pontificalis* contains consular, regnal, and some indiction dates; the inscriptions, we know because Agnellus quotes them, often contain consular, regnal, and/or indiction dates; and so on.

In the entire text, Agnellus mentions no dates at all except those which can be clearly shown to have been taken directly from his sources, as when he quotes an inscription.[37] How, then, did he go about assigning information from different sources to the correct Life of a bishop? The *LPR*'s basic unit of time is the reign of a bishop; as in other historical texts, this *tempus* appears in the phrases which introduce most items, *eo tempore* or *istius temporibus.*[38] Agnellus had no external, consecutive numerical chronology by which to arrange information and events from different sources. He provides no external dates for the reigns of the bishops, all that is given is the length of reign, and in many cases even that information was unavailable. Instead, he

[37] There is evidence for the fact that Agnellus may have personally used indiction dates; toward the end of the work, in ch. 172, he mentions a few contemporary events dated by indictional year.

[38] This technique is not unique to Agnellus; the Roman *Liber pontificalis* uses the same phrases, as does Paul the Deacon's *HL* and the Neapolitan *Gesta*: F. Lanzoni, 'Il "Liber Pontificalis" ravennate', *Rivista di scienze storiche* 6 (1909), 345–70, 425–64, 571–92 (p. 438). In each case, the phrase refers to the reign of a king, bishop or pope.

relied on the name of a bishop of Ravenna being mentioned in each source he used, in order to assign information to the correct Life on his list of bishops.[39] Within Lives also, there is no attempt to present events in chronological order; it is a sort of 'file folder' method of history, with a folder for each bishop, into which contents are placed as discovered.

Many of the sources mentioned above, by their nature, contain the name of the bishop in addition to the date: in particular, inscriptions,[40] dedications, documents. In these cases it would have been obvious to which Life the information belonged, even if Agnellus did not understand the date within the source.[41]

What is perhaps more remarkable is that we can see the same thing going on in Agnellus's use of information from Paul the Deacon's *Historia Langobardorum*, the annal and the Roman *Liber pontificalis*. None of these works is specifically about the bishops of Ravenna, and the passages quoted are not always directly about a bishop. Nevertheless, information is included when it is about a bishop, or when it immediately follows the mention of a bishop in the source.

This can be most easily seen in the use Agnellus made of the *Historia Langobardorum* (hereafter *HL*). Direct paraphrases of several passages, all from Book IV of the *HL* are found in the Lives of Bishops John II, Marinian and John IV.[42] These pieces of information relate either to events in Ravenna or to Byzantine and imperial history during the years 598–610; they do not, however, include every event of this type in the *HL*.[43] Most of the passages

[39] Because bishops with the same name were generally not identified by numbers in this period, it was easy for Agnellus to put information into the wrong Life. This fact goes a long way toward explaining the chronological errors in the *LPR*. It also helps to explain why Agnellus included particular information in an apparently random manner.

[40] Only two inscriptions, and stories associated with them, do not mention the name of a bishop. One of them, in the church of St Michael, contains the date, 'quater pacifici [p.c.] Basilii iunioris uiri clarissimi consulis, indictione .vii.', and is followed by several other similar inscriptions containing the name of Maximian and similar mentions of 'p.c. Basilii iunioris' (*LPR* ch. 77). This information would obviously belong to the Life of Maximian. The other inscription, and the lengthy story which surrounds it, refers to the dedication of the *monasterium* of Sts Stephen, Protasius, and Gervasius next to St Laurence by Lauricius, and includes the date, 'Theodosio XV et Placido Valentiniano'. The Diploma granting metropolitan status to the bishops of Ravenna links the names of Valentinian and Bishop John, as we will see below, and thus an inscription naming Valentinian could also have been connected by Agnellus to John.

[41] Agnellus, *LPR* chs. 48, 51, 57, 60, 66, 70, 72, 74, 75, 77, 85, 92, 93, 112, 114, 115, 170.

[42] Agnellus, *LPR* chs. 98, 101, 102, 106, and 107; Paul the Deacon, *HL* ii. 28–30; iv. 10, 14–15, 17, 20, 25–6, 28–9, 32, 34–6; vi. 49.

[43] A. Testi-Rasponi, 'Note marginali al "Liber pontificalis" di Agnello ravennate', *Atti e memorie della deputazione di storia patria per le province di Romagna*, 3rd s. 27 (1908–9), 86–104, 225–346; 4th s. 1 (1910–11), 397–464 (p. 243) suggests that ch. 94 was written just after Agnellus went to Pavia, and he must have obtained a copy of the *HL* there; also that Agnellus's limited use of the *HL* was based on knowledge of only part of the text. In fact, the events in ch. 94 are the first in which the Lombards began to threaten Ravenna.

from Paul that were directly quoted by Agnellus do not have dates attached to them; their relative chronology in the *HL* is based on the information about various kings and emperors which is given before and after them, a chronological framework which was not being used by Agnellus. However, in Book IV.10, Paul says: 'In that month died John, archbishop of Ravenna, in whose place Marinian, a Roman citizen, was substituted.'[44] The passages quoted by Agnellus from IV. 14–29, up to the death of Gregory the Great, were thus obviously to be placed during the reign of Marinian. Some subsequent events from Paul were placed by Agnellus in the Lives of Marinian's successors, presumably because he felt that some time had passed.[45]

One other passage from the *HL* was quoted extensively by Agnellus, the murder of the Lombard King Alboin by his wife Rosamunda. In the *HL*, the date is given as, 'after three years and six months of Alboin's rule in Italy'.[46] Agnellus tells us that the murder took place 'in the sixth year of the reign of Justin II'.[47] Here, Agnellus was doing some cross-correlation, since for this period he was also relying on his annal, which recorded the murder of Alboin in the sixth consulship of Justin II.[48] So here, Agnellus did not need to use Paul's dates, but combined the story from the *HL* with the annalistic notice. And the annal probably mentioned the bishop Peter III, in whose Life the story was found.

The same methodology can be seen, or inferred, in Agnellus's use of the Roman *Liber pontficalis* and the annals. Agnellus's use of these texts was rather more extensive, and I have explained the arguments in detail

[44] Paul the Deacon, *HL* iv.10: 'eo quoque mense defunctus est Iohannes archiepiscopus Rauennae, cuius in locum Marianus [*sic*], ciuis Romanus substitutus est.'

[45] Another set of quotations, from iv. 14–29, are found in the Life of Marinian's second successor, John IV (625–44). It remains a mystery why he put them into the life of John IV rather than the immediate successor of Marinian, John III, about whom nothing is told except the length of his reign. Agnellus seems to have equated the death of Gregory the Great with that of Marinian, and then put subsequent events somewhat forward in time, because he did not know the length of Marinian's reign. In fact, this information in the Life of John actually did also take place during the reign of Marinian (595–618).

[46] Paul the Deacon, *HL* ii. 28: 'postquam in Italia tres annos et sex menses regnauit . . .' In ii. 7 the Lombards depart from Pannonia in the first indiction, while in ii. 25 Paul says that the conquest of Italy is accomplished by the third indiction. Since other sources put the murder of Alboin in 572, the fourth indiction, Alboin's three-and-a-half year reign must date from the departure from Pannonia, rather than the conquest of Italy.

[47] Agnellus, *LPR* ch. 96; previously he has established that the third or fourth indiction was the fifth year of Justin II, who thus must have acceded to the throne in the fourteenth or fifteenth indiction; the murder of Alboin would have to take place in the fourth or fifth indiction. While Paul does mention (ii. 4) the accession of Justin II, he does not provide a date for this event.

[48] O. Holder-Egger, 'Untersuchungen über einige annalistische Quellen zur Geschichte des fünften und sechsten Jahrhunderts, II, III. Die Ravennater Annalen', *Neues Archiv der Gesellschaft für Altere Deutsche Geschichtskunde* 1 (1876), 309–368 (pp. 339–40).

elsewhere.[49] Here let me simply note that every time a bishop of Ravenna is mentioned in the *Liber pontificalis,* Agnellus provides a corresponding passage, while it seems clear that he took information from the annal based on the mention of a bishop's name at the beginning of a group of entries.

Although sophisticated methods of conversion and synchronistic charts were available to historians of the ninth century, Agnellus does not seem to have used any method of converting between different chronological systems. He relied instead on correlating events in his sources with bishops mentioned by name in them, and organized his text with the reigns of the bishops of Ravenna as the framework. This methodology has significant implications: in particular, dates in this sort of text must be understood as perhaps not well-understood quotations from other sources, rather than points of reference for a larger chronology.

In the *LPR*, the *tempus* is the reign of a bishop. Certainly Agnellus knew that some bishops had much longer reigns than others, but there is no attempt to make the length of the Life correspond to the length of the reign. In a sense, for historical purposes the unit of 'bishop's reign' is a standard one, detached from the number of years it represents.

One implication of this can be seen in the way that Agnellus refers to the passage of time. Agnellus gives only seven references to things happening a number of years ago. He says that he himself was born thirty-two years and ten months previously.[50] Four times he refers to events that took place within living memory, between five and thirty-eight years ago.[51] And twice he offers numbers of years that are round, and probably incorrect. He says that Bishop Martin obtained his seat 'eighty years ago', whereas his reign actually commenced at most thirty-six years before Agnellus was writing.[52] And he says that the amount of time between Apolenaris's day and Agnellus's is '800 years and more',[53] which is certainly roughly accurate, but not based on any numerical reckoning: based on Agnellus's concept of his history, it was actually around 770 years.

I feel that, if other texts are examined in this way, the same sort of pattern

[49] See D. M. Deliyannis, 'The *Liber Pontificalis Ecclesiae Ravennatis*: Critical Edition and Commentary' (unpublished Ph.D. diss., University of Pennsylvania, 1994), pp. 136–47 and 155–70.

[50] Agnellus, *LPR* ch. 54: 'Hodie denique sex gero lustra, duobus insuper annis et bis quinos menses, ex quo lumen recepi . . .'

[51] Agnellus, *LPR* ch. 1: 'Et nune pene annos quinque Theodorus Bononiensis antistes saxeam arcam . . .'; ch. 70: 'usque ad nostra tempora permanserunt, peneque annos 26 demolita sunt . . .'; ch. 94: 'Et nunc pene annis 38, cum Karolus rex Francorum omnia subiugasset regna et Romanorum percepisset a Leone III. papa imperium . . .'; ch. 113: 'Sed pene annos 12, tempore Petronacis pontificis Lotharius augustus tollere iussit . . .'

[52] Agnellus, *LPR* ch. 167: 'pene annos 80 istius adeptus est sedis, consecratus Romae per manus Leonis III. papae . . .'

[53] Agnellus, *LPR* ch. 32: 'a tempore beati Apolenaris post eius decessum pene annos 800 et amplius . . .'

will emerge, namely, authors include dates when they find them in their sources, rather than conscientiously converting all dates to whatever their chronological framework was, if there was one.[54] The correspondence between the types of dates used by historians and those used by notaries in the same place is evident: AD dates and Bede, regnal dates and Gregory, indiction years and Paul. What this means is that modern historians cannot worry about whether or why a medieval historian 'got the date right', since this implies that the medieval historian was thinking of dates as numbers rather than as expressions of authority or antiquarian interest.

CONCLUSION

In the early Middle Ages, time was not generally thought of numerically; rather, year-dates were used for other sorts of validation. In the different areas of Europe, different systems of dating were available. In England, Bede did not introduce AD plus indiction dating, but used it, perhaps because it was already in use on the documents that made up some of his sources. In Francia, regnal dates alone were the standard both on documents and in histories up to the Carolingian period, after which AD dates were adopted for the latter, but only slowly for the documents. In Italy, the Byzantine convention of regnal dating plus the indiction was used both in the Byzantine/papal areas and in the Lombard kingdoms; the indiction, particularly, was important to both notaries and historians. In all cases, if the audience was local, the dating system was local, and this was true in most contexts in the early Middle Ages. If the audience was Christendom, as in ecclesiastical usage, for example, a universal dating system was needed, such as AD, as found on the Easter tables. However, even in ecclesiastical contexts, dating by the indiction was more common than AD until the tenth century.

The dates used by historians correspond to a great extent to the dates found on the documents and other texts that were their primary sources. Except in England, where AD dates were used, the documents were not generally dated according to a universal time-scale. Thus, when historians who were writing longer-term history were confronted with them as sources, they may well have been at something of a loss, unless they too had available the same sort of dating tables used by the notaries. In the case of a complete change of regime, with a concomitant change in dating formulae, the older documents' dates might have been lost to memory, other than knowing that they were 'very old'. Documents dated by consul might have been irrelevant to the Lombards in Italy; documents dated by Merovingian regnal years similarly indeterminate to the Carolingians. Moreover, the dates on foreign

[54] This has been demonstrated by Favrod, 'Les sources' (for Marius of Avenches) and by Humphries, 'Chronicle and Chronology' (for Prosper of Aquitaine).

documents might be totally incomprehensible, as for example, with papal documents, which went all over Europe dated by the reigns of Byzantine emperors. While this may have looked impressive, would it have been comprehensible?

Where this becomes particularly significant is in the marking of the passage of time. One function of a continuous numerical system of dates is that one can tell how much time passed between events. Without such dates at regular intervals, the only way to indicate numerical passage of time is to painstakingly add up the years at various points. This is precisely what Gregory of Tours did in four places in the *Historia Francorum*, coming up with the total number of years from the Creation to significant events. But such calculations were laboured, and served grand theological and historiographical purposes rather than allowing every-day time to be counted. From page to page in the *Historia Francorum*, it is difficult to see at a glance how much time is passing, and the same is true in other narrative texts.

Bede is famous, among other things, for being the first historian to use AD dates. However, it should be noted that this is how his sources were dated. That the practice was adopted only slowly by other historians was probably due to their sources. There is a disjunction between chroniclers, who were using eras like *annus mundi*, as well as more local time-frames, to organize their work, and notaries, who were using primarily regnal dates. Certainly to write history, you had to have a list of something at your elbow. But if you were not writing universal history, you do not seem to have required universal dating standards. Narrative historians arranged their chronologies more along the lines of notarial dating, rather than universal, that is, by reigns of kings or bishops. How to make sense of all the other dates not on your list but in your sources probably varied from author to author. Before the adoption of a universally recognized standard in both documents and history writing, such as the AD system, doing historical research was a difficult task. Early medieval historians, even Bede, are often criticized for getting dates wrong; perhaps we should rather marvel that they got any dates right.

2

Living within Sight of the End

R. A. Markus

At the end of the first century AD, John the Seer of Patmos devoted a chapter of his book of Revelation to warning his readers that 'the time is near' (Revelation 1. 3; 22. 6). I shall be trying to deal with the question as to what it meant for Christians in Late Antiquity to believe this message.

For the fact is that most of these Christians, and indeed in one way or another most Christians at most times, did believe something of this sort. Indeed as Paula Fredriksen, the author of some of the finest studies on early Christian apocalyptic and eschatology once remarked: 'Christianity began with the announcement that time and history were about to end.'[1] One might be inclined to think that, in the light of subsequent events, the announcement turned out to be false, or misleading; but the early generations of Christians who had heard John's warning that 'the time is near' stubbornly refused to budge from their belief. John, writing perhaps in the last years of the first century AD, embodied his vision of the end in bizarre, sometimes reassuring, but more often terrifying imagery. The earth would quake and the heavens be moved, the land would be ravaged and there would be rivers of blood and storms of fire but at the end of it all the saints would be avenged and enjoy a thousand years of peace and reign with Christ. But the decades passed and the end was not yet. The early generations of Christians, however, whether they looked to it with terror or with hope, continued to maintain their obdurate belief that the end would come soon. Early in the second century they were warned against scoffers who would come among them in the last days, saying: 'Where is the promise of this coming [predicted by the prophets and the Lord]? For ever since the fathers fell asleep, all things have continued as they were from the beginning of creation' (II Peter 3. 4). So they continued to hope and to expect the end and to expect it soon; but gradually, as 'soon' turned out to be further and further off, they ceased even to ask themselves how soon was 'soon'. Their expectation of an imminent end in defiance of the facts

[1] P. Fredriksen, 'Apocalypse and Redemption in Early Christianity: from John of Patmos to Augustine of Hippo', *Vigiliae Christianae* 45 (1991), 151–83 (p. 151). See her n. 2 on this statement for references. A shortened form of this paper appears in *The Apocalypse in the Middle Ages*, ed. R. K. Emmerson and B. McGinn (Ithaca, 1992), pp. 20–37. See also her 'Tyconius and the End of the World', *Revue des Études Augustiniennes* 28 (1982), 59–75.

had a profound effect on the way all Christians would subsequently understand time.

Christianity saw itself as the fulfilment of the Jewish prophecies; the promised one expected by the Jews at the end of time had come and he would usher in the Kingdom of God. Christians knew that with the coming of Christ the world entered a new age. The definitive end of world history had been anticipated in events that took place in a Roman province in the reigns of Augustus and Tiberius. The long-awaited kingdom was established, inaugurated, but it remained concealed; it would be revealed only when Christ returned in glory, at the very end, to manifest his power and to gather his faithful from the four corners of the world. Everything was new, but nothing had changed: it was all still to come. The end had come, and yet it had been indefinitely postponed. Such a paradox was fundamental to Christian self-understanding. In the imagery of the great New Testament scholar, Oscar Cullmann, Christians had come to live in a strained, dislocated, time-scheme. The end had been displaced from its proper place at the very end, and was anticipated at the mid-point of time, the time of the earthly life of Jesus Christ, an image reflected in our convention of dating events BC and AD, suggestive of Christ as at the mid-point of time. So we, who live between two ends, the one at the mid-point, the other at what Shakespeare called 'the general all-ending day' (*Richard III* III.I.78),[2] live in a strained time-scheme. That is, both a time of fulfilment, of achieved hopes and promises, and yet also of unrealized hope and of waiting, both before the end, and after it.

It took Christian theologians some centuries to resolve the ambiguities engendered by this contradictory state of being. Moreover, the ambiguities were never wholly exorcized and were to return time and again to haunt the Christian consciousness. For another three hundred years even the main-line teaching in the Christian communities continued to show deep uncertainty on this score. Towards the end of the second century Irenaeus of Lyon cited the ancient prophecies about a messianic kingdom to come, and interpreted them as pointing towards a millennial reign of Christ here on earth. In confirmation, Irenaeus referred to the authority of Papias, 'who had heard John and was a companion of Polycarp [of Smyrna]'.[3] Writing a little more than a century later Eusebius, the 'father of Church History', dismissed such speculations with contempt. Papias, he wrote, had been a man of very limited intelligence, incapable of understanding the symbolic meaning of the apostolic writings and it was his spurious authority, based on his proximity to John and to Polycarp, that misled so many of the fathers including Irenaeus.[4] Eusebius's contemporary, the Latin Lactantius, rehearsed the

[2] W. Shakespeare, *The Complete Works of Shakespeare*, ed. P. Alexander (London, 1951), p. 720.

[3] Irenaeus of Lyon, *Adversus haereses*, ed. I. A. Rousseau and L. Doutreleau, 9 vols. (Paris, 1965–82), V.33.3–4.

[4] Eusebius, *Historia ecclesiastica*, ed. G. Bardy, 4 vols., 3rd edn (Paris, 1984–7), III.39.7–13.

ancient prophecies about the end, saw them confirmed by the Sibylline Oracles and noting a margin of uncertainty about the time of their fulfilment (due to variations in the reckoning of the years), thought that at the most we would have to wait no more than two hundred years.[5]

A conflict of opinions such as this between Eusebius and Lactantius was not new. Long before the time of Constantine there had been many Christians, the great Origen among them, who could not take the ancient apocalyptic prophecies at their face value. Such Christians often rejected the authenticity of the book of Revelation, or adopted strategies for its interpretation that enabled them to bypass a literal understanding which envisaged an earthly, corporeal, millennium. Many others, especially in troubled times, and not only fringe groups such as the Montanists, but even Catholic bishops, 'firm in their belief that the end was upon them, urged their flocks to drastic action'.[6] Such ideas were bound to lose ground with the end of persecution and with Christianity becoming the religion of the emperors, of the respectable and the powerful. Even so, and even after the passing of two or three Christian generations who had never known persecution, millenarian ideas were still alive in the years around 400. Pagans and Christians were both apt to read the events of their own days in terms of the fulfilment of prophecies of the end.

Before proceeding further, it is time, however, to distinguish two sets of ideas: first, the idea of the close proximity of the end, which we might agree to call 'apocalypticism' and define as 'a belief that [the] end, however conceived, is imminent, that one lives at the very edge or in the midst of the Last Events'. This we should distinguish from 'millenarianism' (or, in its Greek form, chiliasm), which has to do not with the time but the shape of the end, defined as 'a belief that these final events will usher in a reign of Peace, Justice, and Plenty *here on Earth* and that salvation for the Just will be collective and its rewards experienced while living in the flesh'.[7] These two sets of ideas are often found held together but, in the interests of clarity, it is as well to distinguish them. Thus Augustine, for instance, never imagined that the end of history was close at hand; but as a young priest, he was apt to interpret the biblical Sabbath rest in millenaristic terms, as the reign of Christ with his saints on this earth preceding the end. He had never taken an

[5] Lactantius, *Epitome divinarum institutionum*, ed. M. Perrin (Paris, 1987), VII.24–5.

[6] Fredriksen, 'Apocalypse', p. 154; see, for a general survey, B. E. Daley, *The Hope of the Early Church. A Handbook of Early Christian Eschatology* (Cambridge, 1991).

[7] R. Landes, 'Lest the Millennium be Fulfilled: Apocalyptic Expectations and the Pattern of Western Chronography, 100–800 CE', in *The Use and Abuse of Eschatology in the Middle Ages*, ed. W. Verbeke, D. Verhelst and A. Welkenhuysen (Leuven, 1988), pp. 137–211 (pp. 205–8). Capitals and italics in the original. Landes also defines 'eschatology' and 'apocalyptic millenarianism' here. Definitions on these lines are widely agreed. Various shades of 'millenaristic' beliefs are distinguished by J. Daniélou, 'La typologie millénariste de la semaine dans le christianisme primitif', *Vigiliae Christianae* 2 (1948), 1–16. On the problems of defining apocalyptic, see C. Rowland, *The Open Heaven: A Study of Apocalyptic in Judaism and Early Christianity* (London, 1982), pp. 23–72.

apocalyptic view of his own time, but traces of millenarian ideas have been found in his early works.[8] Even those, however, were soon to disappear. In the famous final chapters of his *Confessiones*, written in his early forties, there is no trace of such ideas. The last books of his *De civitate dei* contain a frontal attack on them, and chiliasm is condemned as a heresy in his catalogue.[9] The biblical references to 'a thousand years' (Revelation 20. 2; II Peter 3. 8; and Psalms 89 [90]. 4) are to be taken as figurative expressions; there is no reason to expect the end to occur at the close of six thousand years, whatever the way the years since the creation are reckoned. And anyway, a 'thousand years' in the language of the Scriptures should not be taken literally.[10] His definitive answer to such speculations is most emphatically stated in his correspondence in 399 with a Dalmatian bishop, Hesychius, who thought that as the Gospel had now reached all nations, we could expect the end and the Last Judgment very soon. Many years later, Augustine would refer to his reply to Hesychius as his last word on the subject: Jesus himself had taught that not even the angels or the Son knew when it would be (Mark 13. 32) and had warned his disciples not even to ask, for it is not for them to know the times and the seasons (Acts 1. 7).[11]

Augustine's condemnation of speculation about the time of the end, and his refusal to allow for a messianic kingdom this side of the end was derived in part from the work of the dissident Donatist theologian Tyconius. His radical agnosticism about the future course of history contrasts sharply with the great wave of apocalyptic and millenarian speculation which seems to have accompanied the resurgence of a charismatic and ascetic brand of Christianity around the year 400. At the height of the official christianization of the Roman Empire in the age of the Theodosian dynasty one would not expect Christians very readily to discern Rome beneath the purple and the scarlet of the apocalyptic whore; but in fact apocalyptic notions were still in the air in the last decades of the fourth century and the beginning of the fifth and not only in dissenting circles and fringe groups such as the Donatist schismatics in North Africa. End-time expectations and countdowns clustered in the years around 400. Augustine knew many people who thought that the end was at hand.[12] His answer to

[8] G. Folliet, 'La typologie du sabbat chez Saint Augustin: son interprétation millénariste entre 389 et 400', *Revue des Études Augustiniennes* 2 (1956), 371–90 is a classic study. G. Bonner attempts, I think unconvincingly, to qualify this view in 'Augustine and Millenarianism', in *The Making of Orthodoxy: Essays in honour of Henry Chadwick*, ed. R. Williams (Cambridge, 1989), pp. 235–69.

[9] Augustine, *Adversus haereses*, CC 46, 8.

[10] Augustine of Hippo, *De civitate dei*, ed. B. Dombart and A. Kalb, 2 vols. CC 147–8 (1955), XX.7.1–2 and 9.1–2.

[11] Augustine, *De civitate dei* XX.5.4. See, for the correspondence, Augustine of Hippo, *Epistolae*, ed. L. Goldbacher, 5 vols. (1895–1923), 197–9.

[12] References in Fredriksen, 'Apocalypse', p. 179, n. 59. We should note Martin of Tours,

them was the answer he had given to bishop Hesychius of Salona: we do not know, and we should not try to.[13]

Augustine had gradually arrived at a view of history in which this present age that we live in, the stretch between the Incarnation and the Second Coming of the Lord, is entirely homogeneous. This is the sixth and final age of the world before the Last Judgment. Within it there are no divisions, no landmarks which have any significance from the point of view of sacred history, that is to say on the map of God's saving work.[14] 'History's timeframe is known only to God; and if the hour of the end is unknowable in principle, it cannot serve to impose a plot on time . . .'.[15] The plot, like the closure, are unknowable to us. Augustine's banishment of millenaristic and apocalyptic ideas is generally reckoned to be a milestone on the way of 'ridding the [Christian] Church' of such beliefs.[16] But, in the event, Augustine's definitive answer to apocalyptic and millenarian ideas proved to be, as definitive answers often turn out to be, less than decisive. Apocalyptic and millenarian ideas are most likely to appeal to a persecuted minority, to the oppressed and the downtrodden, waiting for deliverance and looking to a future happier than their present lot. Their impact is always liable to be subversive, as countless students of such movements from Norman Cohn to Richard Landes keep telling us. And, in line with such expectations, they continued to flourish, despite Augustine and Jerome, for instance among the oppressed and persecuted Catholics in North Africa under Vandal rule. On the same lines, a Spanish chronicler writing probably in 469, in the conviction that the end was only thirteen years away, gave an account of the last years of his province and of the Roman Empire 'with the barbarians as the agents of the Antichrist, raging against the world amidst portents of growing evil and the fulfilment of the Biblical prophecy'.[17]

Such avowed apocalypticism, however, is very rare among early medieval writers. More usually, like Gregory of Tours and Bede, they disapprove of the messianic movements and popular enthusiasms that they report. In the Latin West the triumph of Augustinian scepticism about the end seems to have been almost complete among the learned and the literate. Such enthusiasms seem to have bypassed the Greek East. Whatever the reasons, they failed to

Sulpicius Severus and Hilarianus, the author of a *Libellus de mundi duratione* (or *De cursu temporum*), who, in the year 397, calculated that a hundred and one years remained (*Du cursu temporum* 16–17, *PL* 13, 1104–5) and Ambrose, *Expositio euangelii secundum Lucam*, ed. M. Adriaen, CC 14 X.10; 14; also in *PL* 15.

13 See above, n. 11.

14 On this, see R. A. Markus, *Saeculum: History and Society in the Theology of Saint Augustine* (Cambridge, 1970, 1988), esp. ch. 2.

15 Fredriksen, 'Apocalypse', p. 166.

16 Landes, 'Lest the Millennium', p. 156; and Fredriksen, 'Apocalypse'.

17 R. W. Burgess, *The* Chronicle *of Hydatius and the* Consularia Constantinopolitana: *Two Contemporary Accounts of the Final Years of the Roman Empire* (Oxford, 1993), pp. 9–10. For the apocalyptic signs and chronology, see the end of the *Chronicle*, p. 122.

take root there: from early times there had been reluctance even to accept the book of Revelation as canonical. True, historians have discovered traces of millennial thinking even in the Byzantine world but it is in the Middle Ages of Latin western Europe that they have found their richest quarry. Much historical research and ingenuity have been devoted to tracing what is claimed to be an almost unbroken line of such thought in European Christianity. A thriving industry has grown from the controversies over the year 1000, devoted to sifting the evidence for millennial and apocalyptic movements from the earliest times. The imminence of the year 2000 even stimulated the foundation of a Center for Millennial Studies at Boston University.[18] To be sure, there are scraps of evidence that apocalyptic and millenarian ideas continued to live a shadowy, subterranean kind of existence. We hear of them from writers who were less than kindly disposed towards them. One historian has contrasted the 'official denial of any millenarianism' in the 'orthodox written tradition' with 'an oral [tradition] which continued to transmit the belief in a sabbatical millennium'.[19] We can recognize such half-submerged millenarianism in the popular enthusiasm alluded to by historians such as Gregory of Tours, aroused by charlatans, false Christs, false prophets and miracle-workers of various kinds, who deceived the simple folk awaiting the end, whether it be with hope or with terror.[20] Although on his own calculations the end was still some two hundred years off, and although his professed belief was that we cannot know when it shall be, Gregory himself interpreted these false Christs and false prophets, along with the famines, pestilence and signs and wonders in the sky, as signs of the end.[21] Bede, following Augustine faithfully, distanced himself much more carefully from such expectations; despite all his care, however, he found himself accused of holding that the world would last only

[18] Its website (http://www.mille.org) gives access to a wide range of bibliographical information and to publications, especially by R. Landes. I wish to thank Dr Claire Taylor for navigating through the Internet and saving me from drowning in its murky depths.

[19] Landes, 'Lest the Millennium', p. 167.

[20] 'eos qui adpropinquantem finem mundi disperant', Gregory of Tours, *Historiarum libri decem*, ed. R. Buchner (Berlin, 1955), I. *Praefatio*, p. 6. The ambiguity of the phrase is noted by G. de Nie, 'Roses in January: a Neglected Dimension in Gregory of Tours' *Historiae*', *Journal of Medieval History* 5 (1979), 259–89 (pp. 281–2). P. Brown has suggested, and justified, the translation 'who do not heed' ('qui ne se souciaient guère') the end of the world for 'qui disperant' in 'Vers la naissance du purgatoire: amnistie et pénitence dans le christianisme occidental de l'antiquité tardive au Moyen Age', *Annales* 52 (1997), 1247–61 (p. 1255, n. 27). See also Gregory, *Historiae* IX.6 and X.25 on charlatans. M. Rouche, *L'aquitaine des wisigoths aux arabes, 418–781* (Paris, 1979), pp. 409–14 discusses the movements of the late sixth century.

[21] Gregory of Tours, *Historiae* X.25; see, for the chronological reckoning, X.31. Landes, 'Lest the Millennium', pp. 166–9 discusses this and notes 'this conflict between the implicit eschatology of chronological calculations and the explicit denials which accompanied them . . .' (p. 166).

six thousand years. What is interesting about the indignant letter he wrote in protest is that it reveals that such a belief was common among *rustici*, who kept pestering him to tell them how many years were left.[22] Against such 'heretical and inane' views (*quia heretica sunt et frivola*), Bede simply asserts that we cannot know: in the rest of the sixth age (this age in which we now live) nothing is certain except that 'this present decrepit age will come to an end with the ending of all history'.[23] 'What is left of the sixth age is known only to God': so ends the shorter of his chronicles.[24]

Occasional isolated allusions of this kind reveal an oral transmission at variance with and running alongside the learned written tradition of the professionals.[25] The trouble is that such allusions are very rare indeed, at any rate in the six hundred years between the time of Jerome and Augustine and the end of the first millennium. The overt evidence for a continuous apocalyptic or millenarian tradition is simply lacking. But, we are urged, we should not allow the silence of the texts to deceive us: 'the textual silence hardly mirrors the actual situation.'[26] The attempt to fill in the gaps and to trace a continuous thread relies heavily on the evidence of the chronological systems used by the annalists and chroniclers. As the eschatological 'count-down' neared the six thousandth year, so this argument runs, the computists simply adopted new systems of reckoning the years, designed to postpone the completion of the world's six thousand year span. Much has thus been inferred from the series of major readjustments of the dating systems culminating in Bede's adoption of the *annus domini* in place of the previously used systems of the *annus mundi*. We are told that:

> the consistent pattern of a long-used dating system disappearing just decades before its apocalyptic term . . . reflects not the absence of millenarianism but its opposite: the strenuous ecclesiastical attempt to combat that belief.[27]

This claim seems to me highly implausible. Such readjustments of time-schemes do not, by themselves, tell us anything about expectations of the end, any more than does my getting a new diary for next year when my

[22] Bede, *Epistola ad Pleguinam de aetatibus seculi*, ed. C. W. Jones, CC 123C 624 (1980), 14 and 15. See also Bede, *De temporum ratione*, ed. C. W. Jones, CC 123B (1977), 67.

[23] Bede, *De temporum ratione* 66.7; Bede, *De temporibus*, ed. C. W. Jones, CC 123C (1980), 16.8 and Bede, *Epistola ad Pleguinam* 4.

[24] Bede, *De temporibus* 22. See also Isidore of Seville, *Etymologiae*, ed. W. M. Lindsey, 2 vols. (Oxford, 1911), V.xxxix.42 and Julianus Toletanus, *De comprobatione sextae aetatis* III.10. *PL* 96, 1260. I discussed these eschatological passages in *Bede and the Tradition of Ecclesiastical Historiography* (Jarrow, 1976), pp. 14–15, repr. in R. A. Markus, *From Augustine to Gregory the Great. History and Christianity in Late Antiquity* (London, 1983).

[25] Landes, 'Lest the Millennium', has collected some of these allusions: e.g. from Ambrosius Autpertus, p. 180.

[26] Landes, 'Lest the Millennium', p. 178.

[27] Landes, 'Lest the Millennium', p. 200.

present one expires. Quite the contrary, what would convince me that an annalist was expecting the world to end as it was approaching the six thousandth year of his calendar would be if he did *not* adopt a new time-scheme: it would not be needed, any more than I would need a new diary if I thought the world would end when it runs out. I buy a new diary, and the annalist a new *annus mundi*, precisely because we expect everything to go on, everything except our calendars. I must confess to a predisposition to believe writers when they say they do not know the time of the end and condemn attempts to calculate it. If I am right, then the use of dating systems as evidence for end-time expectations is wholly inappropriate. They imply nothing unless supported by indications of expectations, whether by *rustici* or others, such as Bede happens to give us. I would like, therefore, to suggest a rule of interpretation which may seem somewhat simple-minded. Let us believe Hydatius, or Gregory of Tours, or Bede, or whoever it is, when he expresses hope or fear of an approaching end; let us believe him when he says that others show such fear or such hope. On the other hand, let us also believe him when he affirms that we cannot know the time of the end, and that we should not seek to know it. And let us not think that he is contradicting himself when he is using a time-scheme which would apparently commit him to thinking the end is X years off, or that when he abandons it, he does so because he is apprehensive that the end would not have arrived when it reaches its end. Let us, rather, think that he is unlikely to be contradicting himself and let us try to understand his meaning.

If we follow these simple rules, the evidence suggests something rather less than a continuous tradition of millenarian expectation faced with 'strenuous ecclesiastical attempt[s] to combat that belief'.[28] To be sure, there are occasional manifestations of such expectations, either in the work of a particular writer, or among the rustics whose views he allows us to glimpse; but they fail to support an almost unbroken tradition of such views running alongside an 'official' ecclesiastical orthodoxy anxious to repress them. It seems to me that a better reading of all the evidence taken together suggests something quite different, and perhaps something more interesting: it points to the complexity of the Christian view of time, and to the various ways in which the complexity is apt to lead to disastrous simplification of one kind or another. I think there is no example that tells us more about the meaning of an expectation of the end than one which has been largely bypassed by historians interested in evidence for attitudes allegedly 'explicitly proscribed by the most powerful and prestigious figures of the day', that of Pope Gregory the Great.[29]

[28] See n. 27, above.

[29] Landes, 'Lest the Millennium', p. 186. So far as I am aware, the only occasion on which he refers to Gregory is in his 'Millenarismus absconditus: l'historiographie augustinienne du Haut Moyen Age jusqu'à l'an mil', *Le Moyen Age* 98 (1992), 355–77; 99 (1993), 5–26 (p. 374, n. 75): Gregory's letter to king Aethelberht.

His biographer was not exaggerating when he wrote that in all Gregory said and did he was conscious of the imminent Day of Judgment.[30] Gregory thought of himself as a faithful follower of Augustine. Naturally, he could hardly appreciate the subtlety and sophistication of Augustine's thought but it is clear to a careful reader that for all the simplification produced by the profoundly changed intellectual environment, Augustine does stand behind much that Gregory has to say. This is true of his eschatological views. Not having a serious wave of millenarian expectations to contend with, such as Augustine had been faced with, Gregory could be much less inhibited in speaking of an imminent end than his mentor had been. Time and again in his preaching and his writing he returns to the theme.

One example will convey its flavour. Preaching on an Advent Sunday, two days after a storm had destroyed houses and churches in Rome, Gregory saw the apocalyptic signs foretold by Jesus (Luke 21. 25–33) being realized all round him:

> Of these [predicted] things some we see already accomplished, others we expect with terror to come very soon. For we see nation rising against nation, their distress afflicting the lands – we see this in our time now more than we read about it in books. You know how frequently we have heard reports from other parts of the world of countless cities being destroyed by earthquake. Plagues we suffer without relief; we do not yet clearly see the signs in the sun, the moon, the stars, but we gather from the change in the air that these are not far off . . . As so many of the things foretold have already occurred, there is no doubt that the few that still remain will soon follow: for the experience of what has come to pass gives us certainty about what is to come.[31]

One could multiply texts from his preaching and his *Dialogi* in which he sees the dire threat of Lombard swords, the perils afflicting Italy, the miseries of a starving and crumbling city, the collapse of ancient institutions and disintegration of the social fabric, as signs of an imminent end. 'In the land in which we live, the world no longer announces its coming end, but shows it forth.'[32] Faced with the crumbling of the fabric of the material world around him, Gregory was driven to look beyond. Even if we had no Gospel to tell us,

[30] John the Deacon, *Sancti Gregorii papae vita* IV.65 (*PL* 75). For what follows, see my *Gregory the Great and his World* (Cambridge, 1997), pp. 51–67; and for further detail and references see C. Dagens, 'La fin des temps et l'église selon S. Grégoire le grand', *Recherches de Science Religieuse* 58 (1970), 273–88; and R. Manselli, 'L'escatologia di Gregorio Magno', *Ricerche di Storia Religiosa* 1 (1954), 72–88. Daley, *The Hope*, p. 213, describes Gregory's reflections on the approaching end as 'homely, restrained and traditional'.

[31] Gregory the Great, *Homiliae in evangelia* I.1.1; 5 and II.35.1–3, *PL* 76, 1075–1312.

[32] Gregory the Great, *Dialogorum libri IV de miraculis patrum Italicorum*, ed. A. de Vogüé, 3 vols. (Paris, 1978–80), III.38.3. See also his *Homiliae in Hiezechihelem*, ed. M. Adriaen, CC 142, I.9.9; II.6.22–24; *Evangelia* I.17.16; *Registrum epistularum*, ed. D. Norberg, 2 vols. CC 140, 140A, III.29; V.37 and XI.37.

he said in a sermon, the world itself proclaims the truth, its ruins are its voice, warning us not to love it;[33] if we still love such a world, 'it is wounds we love, not delights'.[34]

Gregory's message is renunciation. The martyrs, he said, trampled on a world in its flower: 'life was long, people enjoyed good health, material riches, fecundity in procreation, tranquillity in lasting peace; and yet, even while it was flourishing, in their hearts the world was barren.' But it is not so for us:

> Now the world is a barren desert, and yet it flourishes in our hearts. Everywhere there is death, mourning, desolation; we are struck on every side, on every side we are filled with anguish; and yet with blind minds we love the bitterness of the things our flesh desires. We pursue what is fleeting, we cling to what is failing. And because we cannot hold on to what is failing, we fail along with it, as we hold on to it while it crumbles. Once the world held us with its delight; now it is so full of woes, that the world itself calls us to God.[35]

Gregory very often links such observation of the world's decay with warnings to his audience against too close an attachment to a world already approaching its grave. He has turned Augustine's complex and differentiated attitude to the secular world, rooted in his eschatological perspective, into something else: a simple call – intellectually simple if spiritually demanding – for holiness. Many of Gregory's eschatological purple passages serve this end: to proclaim an ascetic Christian morality of renunciation. But this is not always the main burden of his exhortation.

Gregory follows Augustine faithfully in disclaiming any knowledge of 'the times and the seasons';[36] the biblical thousand years are this last, the sixth age, in which Satan remains bound, the Church is built, the Christian life tested, the final Judgment awaited.[37] And yet, the conviction that we live near the end, in the old age of the world, runs like a ground-bass through Gregory's preaching.[38] Gregory was instinctively inclined to read his experience in eschatological terms; especially when it involved what he saw as moral crises. Painful conflict with the emperor or the patriarch of Constantinople, lapses by his fellow clergy which he saw as betrayal of their ministerial calling, and moral failures of all kinds invariably brought on a sense of the closeness of the end. The troubles all around him are the signs

[33] Gregory, *Evangelia* I.4.2.

[34] Gregory, *Hiezechihelem* II.6.22.

[35] Gregory, *Evangelia* II.28.3 is the finest expression of this.

[36] Gregory the Great, *Moralia in Job*, ed. M. Adriaen, 3 vols. CC 143, 143A, 143B, *Praefatio*.10.21 and XXXIV.23.55.

[37] Gregory, *Moralia* IV.9.16; IX.3.5; XVIII.42.67 and XXXII.15.22.

[38] A fine example is Gregory, *Evangelia* I.1. The whole homily (for Advent) is devoted to the theme.

and prodigies spoken of in the Gospel (Luke 21. 25–33). The end is very near; it is the eleventh hour, there is no time to lose and much unfinished business.[39] As he wished the English king Aethelberht to know:

> the end of the world is at hand [*approprinquante . . . mundi termino*] as we learn from the words of Almighty God in the scriptures, and the reign of the saints which has no end is near. As the end approaches, many things threaten which have not happened before: changes in the air, terrors in the heavens, unseasonable tempests, war, famine, pestilence, earthquakes all over the place. They have not all happened in our days, but they will surely follow after our days.[40]

But, Gregory tells the king, do not worry if you see these things; for 'such signs of the end are sent well in advance, so that we might attend to the good of our souls, be mindful of the hour of our death and prepared by our good works for the coming of the Judge.'[41]

The intensity and urgency of Gregory's preaching are easy to misunderstand. The logic of his exhortation is clear, if we attend carefully to what he is saying. What we experience as signs of the end is anticipation of the end which will come 'after our days', nobody knows when, like a thief in the night. We see the signs, now; we expect the end, soon, but who knows how soon? Gregory regularly combines the urgency of imminence with indefinite postponement. What he wants to banish is what he calls *securitas*, which I would like to take the liberty to translate as 'complacency'; this is the point of the temporal separation he makes between the signs and the end.[42] Gregory seems to want his hearers to see their experience in terms of signs of the end, so as to deprive them of the pretext for procrastination, for trivializing responsibilities, to give actions their proper significance in their eschatological perspective. Characteristically, to take just one example, in admonishing a congregation on its duties when electing a suitable person to be their bishop, he would remind them to 'consider the coming day of the eternal judge attentively and to prepare for it in penitence'.[43] Gregory liked to tell the story of a vision in which bishop Cassius of Narni was warned that he would die on the feast of the Apostles, and told by the Lord not to stop doing his work, but to get on with it (*age quod agis*); he was, however, not told the year he would die. As it turned out, the bishop was kept busy for another

[39] Gregory, *Evangelia* I.4.5: 'cum velocitate tempora fugiunt . . .' and I.19.2 – it is the eleventh hour.

[40] Gregory, *Epistularum* XI.37.

[41] Gregory, *Epistularum* XI.37.

[42] On *securitas*, see Gregory, *Epistularum* VII.22; *Moralia* XXIV.11.27; X.20.37 and Gregory, *Evangelia* II.35.1. See, on the temporal disjunction between sign and end, Gregory, *Evangelia* I.1.1–2: 'sequentium rerum certitudo est praeteritarum exhibitio' and II.35.1. See also *inter alia*, Gregory, *Moralia* XX.5.13; Gregory *Evangelia* I.19.2 and II.31.2.

[43] Gregory, *Epistularum* III.29; III.61; IV.23 and IV.44.

seven years, working and preparing for death. *Age quod agis*: it could well have been chosen by Gregory as his own motto.[44]

Gregory's belief in the imminence of the end has nothing to do with calendrical calculations, which he repudiated emphatically. Still less, obviously, is it a 'strenuous ecclesiastical attempt' to combat the belief in the proximity of the end[45] – quite the contrary. What it expresses is his sense of urgency and the conviction that his world could not be taken for granted. Its very instability and the fragility of civilized order demand constant effort of imagination, of understanding and enterprise. He knew that we are constantly tempted to evade responsibility, apt to trivialize through insensibility and lethargy (*hebetudo*) or sheer frivolity. Moral seriousness and urgency, refusal to be defeated by indifference, despair or boredom: these are the things his reminders of the proximity of the end were intended to encourage. And that, in my view, is the core of the Christian understanding of time and its end – a view fatally easy to misunderstand, to simplify and to deform, as it was misunderstood, simplified, and deformed, abundantly in the course of its history. Rightly understood, it is perhaps no less pertinent at the end of our bloodstained century than it was in Gregory's time.[46]

[44] Gregory, *Evangelia* II.37.9. Gregory, *Dialogorum*. IV.58 is taken as his motto by Pope John XXIII.

[45] See above, n. 27.

[46] For a theological study, I would refer to K. Rahner, 'The Hermeneutics of Eschatological Assertions', in *Theological Investigations* 4 (London, 1966), pp. 323–46. See, for example, his description of the future as 'the mystery of God permanently close, menacing and promising . . .' (p. 330); 'An eschatological assertion is not an additional supplementary statement appended to an assertion about the present and the past of man but an inner moment of this self-understanding . . .' (p. 332). In another language, K. Pollmann has recently interpreted Augustine's views in a similar way. See her 'Moulding the Present: Apocalyptic as Hermeneutics in City of God 21–22', in *History, Apocalyptic and the Secular Imagination*, ed. M. Vessey, K. Pollmann and A. D. Fitzgerald (Bowling Green, OH, 1999), pp. 165–81: '. . . Augustine reminds his readers . . . that these are not noncommital images of a world far beyond, but closely linked with the moral and religious life of every individual' (p. 175).

3

Death, Memory and Time: A Consideration of the Mortuary Practices at Sutton Hoo

Howard Williams

Archaeologists, like historians, are concerned with the passage of time.[1] The adequate understanding of the past requires the construction of chronological sequences for sites and regions by a variety of methods, including material culture typologies, seriation techniques, stratigraphy, radiocarbon dating and dendrochronology.[2] In combination, these methods allow archaeologists to identify when, how and possibly why things changed. Yet archaeologists have also been interested in the subjective experience of time. They are concerned with how time and social memories are constructed through social practice.[3] There is a need to consider the ways by which objects, graves and monuments revealed by archaeological research operated in different spheres of temporality and were used to articulate and maintain links with the past.[4] For the early medieval period in Britain, this subject offers a fascinating opportunity to reconsider the archaeological evidence for burial sites.

In recent years, most discussions of how early medieval people perceived their own past have focused upon the use of the written word. Historians have scrutinized the social and political contexts within which histories, regnal lists, genealogies and even charters were written, stored, circulated and exchanged, collected and sometimes destroyed.[5] It has been argued that

[1] I would like to thank the following for their comments, assistance and support: Martin Carver, Tania Dickinson, Yannis Hamilakis, Cornelius Holtorf, Chris Humphrey, Andrew Jones, Marion Manwaring, Aliki Pantos, Sarah Semple and Barbara Yorke. All errors are my own responsibility.

[2] For an introduction to chronological studies in early medieval cemeteries, see *The Pace of Change: Studies in Early-Medieval Chronology*, ed. J. Hines, K. Hoilund Nielsen and F. Siegmund (Oxford, 1998).

[3] M. Shanks and C. Tilley, *Social Theory and Archaeology* (Cambridge, 1987), pp. 118–36; C. Gosden, *Social Being and Time* (Oxford, 1994).

[4] D. Lowenthal, *The Past is a Foreign Country* (Cambridge, 1985); *Who Needs the Past? Indigenous Values and Archaeology*, ed., R. Layton (London, 1994); E. Hobsbawm and T. Ranger, *The Invention of Tradition* (Cambridge, 1983); *The Past in the Past*, ed. R. J. Bradley and H. Williams, World Archaeology 30 (1) (London, 1998); C. J. Scull and A. F. Harding, 'Two Early Cemeteries at Milfield, Northumberland', *Durham Archaeological Journal* 6 (199), 1–29.

[5] E.g.: P. J. Geary, *Phantoms of Remembrance: Memory and Oblivion at the End of the First*

whether the individuals or events they discuss are 'historical', 'mythical' or a combination of the two, they served to situate and legitimize the identities and social relations of individuals, groups and kingdoms. Oral traditions for centuries may have communicated such social memories, and transformed them in different socio-political contexts, but Christianity and the written word allowed them to achieve a new permanence and to be employed with greater legitimacy. However, this focus upon the written word has its limitations, since, as many have argued, in societies where literacy is absent or restricted to a small group, processes of cultural transmission and remembrance take a myriad of other forms.[6] For early medieval Europe, the complex relationships between literacy and orality in perceptions of the past is one area of discussion.[7] Furthermore, Patrick Geary has noted that a range of resources were used in the construction of social memory alongside the written word. These include the social use of objects, monuments, rituals, places and the landscape itself.[8] For example, Geary shows how books affect memory in a variety of ways, not only by their contents but also by their materiality. We can suggest that, in many early medieval societies, the written and the spoken word are only elements of a wider set of strategies by which societies selectively remember and forget elements of social time and memory. One particular strategy of remembrance discussed at some length by Geary is the relationship between the living and the dead.[9] His perspective is a literary one, but from an archaeological point of view this opens many possibilities. Graves dated to the fifth to seventh centuries AD in England are common archaeological discoveries, so if memories could be transmitted through media other than the written word, burials may themselves inform us about the ways early medieval communities constructed social memories.

Millennium (Princeton, 1994); N. Howe, *Migration and Mythmaking in Anglo-Saxon England* (New Haven, 1989); M. Innes 'Memory, Orality and Literacy in an Early Medieval Society', *Past and Present* 158 (1998), 3–36; H. Wolfram, '*Origo et religio.* Ethnic Traditions and Literature in Early Medieval Texts', *Early Medieval Europe* 3 (1994), 19–38; M. T. Clanchy, *From Memory to Written Record: England, 1066–1307*, 2nd edn (Oxford, 1993); S. Foot, 'Remembering, Forgetting and Inventing: Attitudes to the Past in England at the End of the First Viking Age', *Transactions of the Royal Historical Society*, 6th s. 9 (1999), 185–200; P. J. Geary, 'Land, Language and Memory in Europe, 700–1100', *Transactions of the Royal Historical Society*, 6th s. 9 (1999), 169–84; C. Cubitt, 'Monastic Memory and Identity in Early Anglo-Saxon England', in *Social Identity in Early Medieval Britain*, ed. A. Tyrell and W. Frazer (Leicester, 2000), pp. 253–76.

[6] M. Rowlands 'The Role of Memory in the Transmission of Culture', *World Archaeology* 25 (1993), 141–51.

[7] E.g. Innes, 'Memory', pp. 3–6.

[8] Geary, *Phantoms of Remembrance*, pp. 124–8.

[9] Geary, *Phantoms of Remembrance*, pp. 48–80; P. Geary *Living with the Dead in the Middle Ages* (Ithaca, 1994).

SOCIAL PRACTICE, DEATH AND MEMORY

Before discussing the burial archaeology of early Anglo-Saxon England, we need to consider an important question. How might mortuary practices aid in the process of remembrance; both remembering and forgetting? Paul Connerton has argued that social practices have an important part to play in 'how societies remember'.[10] The focus is not on individual memory, but on how shared views of the past are created that inform the identity of a particular community or social group. Connerton makes a broad distinction between what he sees as two strategies of remembrance: inscribing practices and incorporating practices.[11] The former, which includes written and archival histories as well as iconographic and monumental modes of remembrance, provides for the storing of facts in a variety of media. However, Connerton's study emphasizes the importance of incorporating practices of remembrance in societies past and present. These include a range of practical actions, rituals and performances focused upon somatic experience. Inscribing and incorporating practices are not mutually exclusive categories; but by emphasizing the location of memory within social relationships and group rituals, rather than upon a storage on paper or in the mind, Connerton follows Halbwachs in regarding collective or social memories as created by and transmitted between people through social practices.[12] Memory is a form of work.[13] Using case studies from recent European history, Connerton identifies examples of 'commemorative ceremonies' in the creation of time and history through the depiction of past events in rituals.[14] His diverse examples of the importance of commemorative and public performances for creating social memory include the parades of Hitler's National Socialist Party in 1930s Germany and the Christian liturgical calendar. Rituals serve to link past and present; and through their regular occurrence, form and content, they serve to emphasize certain types of remembrance over others. Connerton also shows how 'bodily practices' can be employed in strategies of remembrance: that is, habits and techniques of the body provide means by which the past is reproduced in the present.[15] In the case of the French court of the eighteenth century, he argues that social memories of social standing, genealogies and familial history were

[10] P. Connerton, *How Societies Remember* (Cambridge, 1989). For a brief review of this approach and its archaeological and anthropological application, see Rowlands, 'Role of Memory'.

[11] Connerton, *How Societies Remember*, pp. 72–3.

[12] Connerton, *How Societies Remember*, pp. 36–40.

[13] Rowlands, 'Role of Memory', p. 144.

[14] Connerton, *How Societies Remember*, pp. 41–71.

[15] Connerton, *How Societies Remember*, pp. 72–104.

articulated through non-verbal, non-written media, including daily routines and manners of dressing, walking and talking.[16]

Connerton does not specifically discuss the commemorative and bodily practices that surround death in many societies, but his work has been influential for anthropologists studying mortuary practices (see below). It also opens up numerous possibilities for archaeologists to assess the importance of the material evidence of burial sites in the production and reproduction of social memories.

It can be argued that mortuary practices in many societies may have an active role in mediating the physical dissolution of the social person and, through the rite of passage, ensuring their reconstitution and re-membering.[17] Let us now explore further how different elements of the mortuary rites may influence the way societies remember their dead and the past, using case studies from the anthropological and sociological literature. Throughout, it is important to stress that social memory in mortuary contexts often focuses as much on forgetting (the suppression of the past) as on the conscious evocation of the dead and their identities.

Monuments, memorials, images and memory

First, let us take monuments to the dead. Monuments endure, survive beyond the corpse, and hence it is easy to think of them as implicated in the process of remembrance. However, we must be careful not to regard the mnemonic role of monuments in a simple way. Rather than seeing monuments as stores of information, we might imagine them working in more complex ways through both inscribing and incorporating practices of remembrance.[18] First, their exact form will selectively enhance certain kinds of memory and temporality over others. Furthermore, monuments do not simply 'preserve' memories. In Maurice Bloch's well-known study of the Merina we see an example in which substantial, monumental tombs were prepared for the dead, but the aim was in no way to 'preserve' the memory of deceased individuals, but to serve as a collective focus for an ancestral community.[19] Hence, monuments embody both remembering and forgetting and the conceptual placing of the newly dead in relation to hazy ancestral or collective identities. In other contexts, monuments are less concerned with retrospection and more with prospection: that is, providing an image of the future identity and memory of the deceased, perhaps in the context of an afterlife.

[16] Connerton, *How Societies Remember*, pp. 79–88.

[17] Bloch, 'The Past and the Present'.

[18] C. Holtorf, 'Megaliths, Monumentality and Memory', *Archaeological Review from Cambridge* 14 (1996), 45–66; S. Tarlow, *Bereavement and Commemoration: An Archaeology of Mortality* (Cambridge, 1999).

[19] M. Bloch, *Placing the Dead* (Prospect Heights, IL, 1971).

In some societies, images and portrayals of the dead, as well as mentioning their names, are strictly forbidden, serving to facilitate the 'forgetting' of the dead.[20] In other societies, the clear display of images also serves this purpose. As Siegel notes concerning the widespread use of photographs in Indonesian funerals: 'Javanese funeral photography . . . operates against memory.' Even though the dead are portrayed, they are rendered timeless and static – a denial of the dissolution and decomposition that the body undergoes.[21] The photographs serve to encourage the forgetting of the dead, in the same way that the communal tombs of the Merina serve in the 'forgetting' of the individual social persona as part of the secondary burial rites. Monuments, images and epitaphs have a complex practical involvement in the process of remembrance.

We can find another challenge to the view that monuments are simply aids to memory in the monuments of the First World War. The stylized portrayal of the war dead does not so much remember acts as transform the dead into a new context. In this sense, as Jay Winter argues, war memorials are concerned with forgetting as much as remembering.[22] Winter also notes that their design is often antiquated, as if in order to situate them in the context of a long history of martial activity.[23] Rowlands observes a similar rooting of the imagery of war memorials in a 'primordial past'.[24] Furthermore, monuments are not solely concerned with inscribing memory on place: they also create a context for commemorative and bodily practices. War memorials become the focus of a range of commemorative and bodily practices that change over time, whereby the past is reworked to suit the needs of the present community. Their physical presence serves to influence the spatial movement, orientation and interaction of people in a variety of ways: for instance, by creating zones of exclusion and inclusion.[25] It is the repeated rituals at these sites that implicate them in social memory, not simply because memory is inscribed by their physical presence. So, monuments are implicated in memory reproduction in a variety of ways, 'inscribed' on them in a manner analogous to a text, but 'incorporated' by influencing the spatial movement and the bodily experience of agents.

[20] A. C. Taylor, 'Remembering to Forget: Identity, Mourning and Memory among the Jivaro', *Journal of the Royal Anthropological Institute Man* 28 (1993), 653–78.

[21] J. T. Siegel, 'Images and Odors in Javanese Practices Surrounding Death', *Indonesia* 36 (1983), 1–14.

[22] J. Winter, *Sites of Memory, Sites of Mourning: The Great War in European Culture and History* (Cambridge, 1995), p. 115. See also Tarlow, *Bereavement*, pp. 147–70.

[23] Winter, *Sites of Memory*, p. 84.

[24] Rowlands, 'Role of Memory', p. 146.

[25] Winter, *Sites of Memory*, pp. 102–5; A. King, *Memorials of the Great War in Britain: The Symbolism and Politics of Remembrance* (Oxford, 1998), pp. 128–72.

Objects, feasting and memory

So far, we have considered the roles of enduring structures in the remembrance of the past; but now I would like briefly to consider the roles of portable objects: tools, clothing, vessels, foodstuffs and even animals. Objects survive the dead, and can act to embody and transmit the remembrance of the past in a variety of ways. Histories can be inscribed on objects, and henceforth they carry these biographies with them through circulation and exchanges. Like monuments, images and statues, more portable forms of material culture can also have histories inscribed on them, and their antiquity and circulation can give them 'personalities' of their own.[26] Memories can also, however, be reproduced through the destruction, impermanence and incorporation of portable items as well as by their curation.

Ethnographies of Melanesia have addressed the way in which gift exchanges and feasting during mortuary practices can be important means of structuring the remembering and forgetting of the dead and the social relationships in which they were a part. Battaglia's work among the Sabarl is a clear example of how the collection and distribution of wealth serves to influence the forgetting of the dead and the debts they owed. She states quite simply that 'gift giving creates memory', and these actions have greater precedent over the spoken word:[27] 'In contrast to words, which in the local view are unreliable and unstable vehicles of memory, material wealth and the gift exchange performances that incorporate it are highly valued as "memories" and they are very closely associated with their donors.'[28] Such performances honour and produce fame for the deceased and survivors and will be remembered for a long time, yet the identity of the deceased is suffused within these communal events.

A contrasting relationship between objects and memory is described by Susanne Küchler in her work on the wooden sculptures called Malangan from New Ireland. These elaborate objects are destroyed at the final stage of the mortuary rites, but the memory of their image is thus released from its physical form and can be reused in future mortuary practices. 'As gifts, ephemeral objects confound time not through permanence, but through renewal and are thus central to the production of memory in culture and society.'[29] So in this case, it is the destruction of objects that characterizes the means by which social memory is reproduced.

[26] Rowlands, 'Role of Memory', p. 149.

[27] D. Battaglia, 'The Body in the Gift: Memory and Forgetting in Sabarl Mortuary Exchange', *American Ethnology* 19 (1992), 3–18.

[28] Battaglia, 'The Body in the Gift'.

[29] S. Küchler, 'Malangan: Art and Memory in a Melanesian Society', *Man* 22 (1987), 238–55; S. Küchler, 'Malangan: Objects, Sacrifice and the Production of Memory', *American Ethnologist* 15 (1986), 625–37; S. Küchler 'Landscape as Memory: The Mapping of Process and its Representation in a Melanesian Society', in *Landscape: Politics and Perspectives*, ed. B. Bender (Oxford, 1993), pp. 85–106.

The work of both Küchler and Battaglia looks at the way the production, display and destruction of objects stand for, and influence, the remembering of the dead. Richard Eves has emphasized the importance of eating and drinking in the working of memory in the mortuary context, since feasting affects remembrance through (and incorporated into) the body.[30] In his study of the Lelet of New Ireland, mortuary feasts are created to promote some memories over others. Remembering can take place through the body, but also the body represents a site where memories are constituted. The cycle of mortuary feasts serves to 'finish' the dead – to instigate the social forgetting of the individual. The construction, revelation and destruction of a Malangan figure serves to articulate this process of forgetting. But feasts not only memorialize the dead: they also promote the host and their clan.[31] Eves emphasizes the fact that, while certain memories are circulated and promoted in the mortuary rites, others are erased through this same process. He also suggests that mortuary rites are not simply about finishing the dead, but are contexts in which new memories and social relationships are created. 'Mortuary feasts are enacted events of extreme cultural importance, events which . . . are centrally concerned with memory, both remembering and forgetting, as an intertwined mnemonic process.'[32]

There are numerous other anthropological studies of the destruction of property as a means of decomposing and reconstituting the identity of the deceased through mortuary rites, and it is to the treatment of the dead body itself that we must next turn.[33] What is clear is that objects are related to social memory in more complex ways that might be at first supposed.

The body and memory

So far we have discussed objects, monuments and images employed in mortuary practices. What about the mnemonics of the body itself? Connerton argues that the body is the focus of memory, that through the actions of the body, the past is 'sedimented' on to it; but as mentioned above, we should extend this to the body during and after death and the associated rituals and ceremonies.[34] The dead body is often the focus of powerful emotional and

[30] R. Eves, 'Remembrance of Things Passed: Memory, Body and the Politics of Feasting in New Ireland, Papua New Guinea', *Oceania* 66 (1996), 266–77. See also R. J. Foster, 'Nurture and Force-feeding: Mortuary Feasting and the Construction of Collective Individuals in a New Ireland Society', *American Ethnologist* 17 (1988), 431–48.

[31] Eves, 'Remembrance of Things Passed', pp. 269–70.

[32] Eves, 'Remembrance of Things Passed', p. 271.

[33] S. Kan, *Symbolic Immortality: The Tlingit Potlatch in the Nineteenth Century* (Washington, DC, 1989); J. Okely, *The Traveller-Gypsies* (Cambridge, 1983).

[34] Connerton, *How Societies Remember*, p. 72. For death and the body, see C. Schilling, *The Body and Social Theory* (London, 1993), pp. 175–98.

painful somatic responses, and through its decompositions (or attempts to avoid them), it produces a range of forceful affects on mourners and participants.[35] Following Hertz, we can see the washing, clothing, transportation and disposal of the body acting recursively with the imagined transformation of non-corporeal elements of the deceased and the survivors.[36] These treatments simultaneously serve to reconfigure the relationships between the living and the dead and hence the processes of remembrance. Material culture has a central role in this process, especially for objects closely connected with the body and undergoing the same transformations. Brad Weiss has provided a case where the shroud has a central role in the working of memory among the East African Haya of Tanzania. As an inalienable object, it serves to maintain relations between the living and the dead while symbolically distancing and forgetting the distinctive identity of the deceased.[37]

The transformation of the body through technological processes such as excarnation, mummification or cremation, or the movement of the body during funerary processions, may serve to articulate the changing identity of the deceased and mourners. In north Indian Hindu rites, the body is not only changed in state by cremation, but also through post-cremation rites it takes on new bodies such as the symbolizing of the *preta* (spirit) in the form of the *pinda* (ball of rice) and the scattering of the ashes in a sacred river such as the Ganges. This re-presentation of the body following its dissolution has a fundamental impact on the nature of remembrance and gradually situates the dead within the context of social relations and Hindu cosmology.[38] Equally, the secondary burial of human remains in a communal tomb, known famously from the Merina of Madagascar, serves not only to establish a community of ancestors, but to actively embody in place the forgetting of the living individual.[39] Piers Vitebksy's study of death and shamanism among the Sora of Orissa makes the transformation of memory a central concern. He charts how the funeral and shamanistic techniques serve to mediate the transformation of the deceased's memory or *sonum*. The dead are slowly forgotten, transformed in social, cosmological and ontological status into different places in the landscape surrounding the village by both the 'commemorative ceremonies' of the cremation and post-cremation rites, but

35 L. Connor, 'The Action of the Body on Society: Washing a Corpse in Bali', *Journal of the Royal Anthropological Institute* 1 (1995), 537–59.

36 R. Hertz, 'A Contribution to the Study of the Collective Representation of Death', in *Death and the Right Hand*, trans. R. Needham (London, 1960); P. Metcalf and R. Huntingdon, *Celebrations of Death*, 2nd edn (Cambridge, 1991); A. Van Gennep, *Rites of Passage* (London, 1960).

37 B. Weiss, 'Forgetting Your Dead: Alienable and Inalienable Objects in Northwest Tanzania', *Anthropology Quarterly* 70 (1997), 164–72.

38 J. Parry, *Death in Benaras* (Cambridge, 1994).

39 Bloch, 'The Past and the Present'.

also by the 'bodily practices' of the shamanic séances.[40] Such studies emphasize the role not only of memorial but also of technological processes and ritual performances concerning the body in the process of remembrance.

Space, place and landscape

Finally, we need to consider the spatial organization of burial rites in the process of remembrance. Mortuary practices can embody the most traditional elements of societies and the place of burial can become a site of memory where the key values of a group are articulated and manifest. Küchler's work, discussed above, shows how the production and destruction of Malangan figures in mortuary ceremonies of New Ireland is concerned with the remembrance of rights and inheritance of land.[41] Meyer's edited volume on ethnic cemeteries in the United States provides a series of studies of the way in which ethnic groups over the last two centuries have articulated membership to the group through their distinctive burial sites. These works also show how the cemetery can become synonymous with the remembrance of mythical homelands.[42] For Ukrainian communities in the United States, gravestone design seeks to evoke remembrance of their homeland, as well as the deceased and his family.[43] In these ways, burial sites can become more than places of remembrance of individuals by close relatives and friends. Over time, they may become the embodiment of communal myths and histories. Memories are inscribed on place, but also incorporated into people's emotive and experiential relationships with the land.[44]

Exhumation and the politics of memory

Following Hertz, we have already discussed how the funerary process as a rite of passage involves the reconstitution of social memory through practical actions. Secondary burial and 'ancestor rites' can also reaffirm these memories through visiting the grave, memorials or other places associated with

40 P. Vitebsky, *Dialogues with the Dead: The Discussion of Mortality among the Sora of Eastern India* (Cambridge, 1993), pp. 66–93, 176–235.

41 Küchler, 'Malangan'.

42 *Ethnicity and the American Cemetery*, ed. R. E. Meyer (Bowling Green, OH, 1993).

43 T. E. Graves, 'Keeping Ukraine Alive Through Death: Ukrainian-American Gravestones as Cultural Markers', and K. Kiest, 'Czech Cemeteries in Nebraska from 1868: Cultural Imprints on the Prairie', in *Ethnicity and the American Cemetery*, ed. Meyer, pp. 36–76, 77–103.

44 For review of peoples' relationships to place and landscape, see: *The Anthropology of Landscape*, ed. E. Hirsch and M O'Hanlon (Oxford, 1995); C. Tilley, *A Phenomenology of Landscape* (Oxford, 1994); *The Archaeology and Anthropology of Landscape: Shaping Your Landscape*, ed. P. Ucko and R. Layton (London, 1999). For an introduction to the study of landscape context of burial sites, see M. Parker Pearson, *The Archaeology of Death and Burial* (Stroud, 1999), pp. 124–41.

the dead individually or collectively. Katherine Verdery's discussion of post-Soviet Eastern Europe is a prime example of how memories can be transformed, sometimes in an overtly political and ideological context, through a combination of the media discussed above long after the funeral is complete.[45] In particular she is concerned with the ways that material objects and dead bodies form the focus for rewriting histories following the end of the Soviet Union. The ideological change brought about by the end of Soviet Russia saw new and ambivalent attitudes towards public statues and memorials of communist leaders, sometimes leading to their destruction – a process of social forgetting.[46] Equally, individuals returning to favour were exhumed and returned to their country of origin to be reburied with great pomp.[47] Verdery argues that dead bodies act as symbolic capital through which the past is created, narrated and reconfigured. Rather than an Orwellian-style rewriting of history books, she argues that communities and nations in the modern world can rapidly employ the dead to reconstruct and manipulate their history.[48] For example, in 1987 the remains of the fourteenth-century Serbian prince who had been killed fighting the Turks was exhumed and taken on a tour of ancestral Serbian territories before his remains were reburied in Kosovo.[49] This particular example draws together many of the themes discussed above, while showing how their significance for creating social memories extends far beyond the initial disposal of the body. It also demonstrates the explicit political and ideological contexts within which social memory can operate.

Summary

We have reviewed some of the recent literature that addresses how the somatic, technological, material and spatial aspects of mortuary practices are important in both remembering and forgetting the dead. Beyond these general points about ritual and the body, there are a number of specific reasons why mortuary practices might be important contexts for social memory to be reproduced in many societies. These can be summarized in a number of inter-relating points. Mortuary practices are often believed to be rehearsals of sacred archetypes situated in the distant past.[50] Furthermore, funerals involve not only social and personal changes, but also cosmological

[45] K. Verdery, *The Political Lives of Dead Bodies: Reburial and Postsocialist Change* (New York, 1999).

[46] Verdery, *Political Lives of Dead Bodies*, pp. 4–13

[47] Verdery, *Political Lives of Dead Bodies*, pp. 13–20.

[48] Verdery, *Political Lives of Dead Bodies*, pp. 115–27.

[49] Verdery, *Political Lives of Dead Bodies*, p. 18.

[50] M. Eliade, *The Myth of Eternal Return* (Princeton, 1954); M. Eliade, 'Mythologies of Death: An Introduction', in *Religious Encounters with Death. Insights from the History and Anthropology of Religions*, ed. F. Reynolds and E. H. Waugh (London, 1977), pp. 13–23.

and cosmogonic themes. Jonathan Parry's work on cremation in Benaras shows how each cremation is believed to be a cosmogonic event, recreating the fire of Agni that first created the world.[51] Death also creates distinctive sensory conditions, particularly smells, which may create distinctive remembrances. For example, Siegel and Bubandt discuss the ways in which smell mediates memory in Indonesian funerary rites, through the need to avoid the smell of decomposition and the various substances used to hide it.[52] Mortuary practices can stay similar for generations, or change rapidly according to different historical conditions. It is often the case that, whether they change or stay the same, funerary rites are often perceived as preserving the traditions and identities of a community. For instance, Hindu and Sikh immigrants into Britain have had to make radical changes to the sequence and organization of their funerary rites, yet active efforts have been made to regard the ceremonies surrounding death as among the most conservative practices that draw a range of parallels and allusions to their ethnic and religious identities and places of origin.[53] Hence they are often regarded as linkages between past and present through the ritual acts, as well as through the deceased.

ARCHAEOLOGIES OF DEATH AND MEMORY

Archaeologists are increasingly taking on these ideas of social memory and time in discussing the material remains of the past, and burial rites in particular. The role of monuments in the reproduction of social memories has been remarked most frequently in recent years. For instance, Richard Bradley has discussed the ways in which archaeologists may be able to identify the ways and means by which memory and time was reproduced through the phases of Stonehenge.[54] Julian Thomas has developed similar ideas when looking at the construction of identity through the use of the Mount Pleasant henge near Dorchester in Dorset.[55] By their construction and use, monuments embody certain memories and perceptions of time, and

[51] Parry, *Death in Benaras*.

[52] N. Bubandt, 'The Odour of Things: Smell and the Cultural Elaboration of Disgust in Eastern Indonesia', *Ethnos* 63 (1998), 48–80; Küchler, 'Landscape as Memory', p. 99; S. Kus, 'Towards an Archaeology of Body and Soul', in *Representations in Archaeology*, ed. J.-C. Gardin and C. Peebles (Indianapolis, 1992), pp. 168–177 (p. 169); Siegel, 'Images and Odors'; Tarlow, *Bereavement*, pp. 20–49.

[53] S. Firth, *Dying, Death and Bereavement in a British Hindu Community* (London, 1997); S. Singh Kalsi, 'Change and Continuity in the Funerary Rituals of Sikhs in Britain', in *Contemporary Issues in the Sociology of Death, Dying and Disposal*, ed. G. Howarth and P. C. Jupp (London, 1997), pp. 30–43.

[54] R. Bradley, *The Significance of Monuments* (London, 1998), pp. 85–100. See also Parker Pearson, *Archaeology of Death*, pp. 193–7.

[55] J. Thomas, *Time, Culture and Identity* (London, 1996), pp. 183–233.

through their long-term use can come to epitomize the past in the present. Yet, there are also many cases where archaeologists can identify monuments that are appropriated into new webs of significance and new strategies of remembrance long after their construction and initial abandonment. Archaeologists are increasingly interested in the 'afterlife of monuments',[56] charting the 'life-histories' of individual monuments or groups of monuments to identify the ways in which their monumentality evokes different practical uses and responses in varying social, political and economic contexts over time. For instance, Holtorf has discussed the cultural biographies of megalithic tombs in north-east Germany from the Neolithic through to the present day, and the varied ways in which they are altered, transformed and reconstituted both physically and conceptually.[57]

For early medieval Britain and Ireland, the catch-all term 'monument reuse' is used to describe the re-activation of prehistoric and Roman sites for new purposes after substantial periods of little or no activity. In a myriad of ways the reuse of ancient monuments could be implicated in the construction and reproduction of memories because their antiquity could have emphasized specific conceptions of time and history that would be more difficult to achieve if social and ritual practices took place elsewhere.[58] This practice has been most famously discussed for Yeavering, which in its later phases is likely to have been the Northumbrian royal palace of *Ad Gefrin* referred to by Bede.[59] The halls were set out and aligned upon two prehistoric monuments (a barrow and a probable stone circle) that remained visible landmarks despite centuries of abandonment. Furthermore, both monuments were reconstructed and provided with timber post additions as well as forming the focus of burial sites.[60] It may be argued that archaeological evidence such as this bears testimony to the deliberate manipulation of conceptions of time and memory through ritual practices including the burial of the dead.[61]

Archaeologists are realizing that monumentality is not the only com-

[56] R. Bradley, *Altering the Earth* (Edinburgh, 1993).

[57] C. Holtorf 'The Life-histories of Megaliths in Mecklenburg-Vorpommern (Germany)', *World Archaeology* 30 (1998), 23–39. For a landscape perspective, see N. Spencer, 'Heroic Time: Monuments and the Past in Messenia, Southwest Greece', *Oxford Journal of Archaeology* 14 (1995), 277–92.

[58] *The Past in the Past*, ed. Bradley and Williams. See also S. Semple, 'A Fear of the Past: The Place of the Prehistoric Burial Mound in the Ideology of Middle and Later Anglo-Saxon England', *World Archaeology* 30 (1998), 109–26; E. Thäte, 'Alte Denkmäler und frühgeschichtliche Bestattungen: Ein sächsisch-angelsächsicher Totenbrauch und seine Kintinuität', *Archäoligische Informationen* 19 (1996), 105–16; B. Effors, 'Monuments and Memory: Repossessing Ancient Remains in Early Medieval Gaul' (forthcoming).

[59] B. Hope Taylor, *Yeavering: An Anglo-British Centre in Early Northumbria* (London, 1977).

[60] R. Bradley, 'Time Regained: The Creation of Continuity', *Journal of the British Archaeological Association* 140 (1987), 1–17.

[61] H. Williams, 'Monuments and the Past in Early Anglo-Saxon England', *World Archaeology* 30 (1998), 90–109.

ponent of the archaeological record through which memory work can be identified. The detailed use of the burial space can also yield important clues, and John Barrett, Paul Garwood and Koji Mizoguchi have interpreted the development of Bronze Age burial mounds and barrow cemeteries in terms of the construction of community histories and genealogies. They argue that earlier barrows structured and influenced the development of the cemetery, both in the location in which new barrows were placed, and in the ordering and character of the burials.[62] Similarly, Jonathan Last has discussed the complex variability in the orientations, posture, age and sex of burials and their sequence of placement within a Bronze Age barrow from Barnack in Cambridgeshire in terms of memory and attitudes towards the body.[63] Therefore, these studies are looking at the relationship between mortuary practices and memory in more complex ways than simply discussing the erection of monumental 'memorials' to remember the dead either singly or collectively. The mortuary context can serve in the negotiations of sacred memories, social memories, the identity of the deceased and his/her kin in a variety of ways through the treatment of the body, burial of grave goods, spatial relationships with earlier graves or ancient monuments, as well as through monument building and monumental architecture.

In early medieval archaeology, several authors have discussed how objects, through the ritual context of their use, or through the stylistic elements of their design might embody ideologies and myths of origin and identity. For example, Heinrich Härke has suggested that the inclusion of weapons in graves formed part of a Germanic origin myth.[64] Others have discussed how the use of gold bracteates, swords, helmets and animal art formed part of political narratives of mythical origin and elite identity.[65]

[62] J. C. Barrett, 'The Monumentality of Death: The Character of Early Bronze Age Mortuary Mounds in Southern Britain', *World Archaeology* 22 (1990), 179–89; J. C. Barrett, *Fragments from Antiquity: An Archaeology of Social Life in Britain, 2900–1200 BC* (Oxford, 1994), pp. 109–31; P. Garwood, 'Ritual Tradition and the Reconstitution of Society', in *Sacred and Profane*, ed. P. Garwood (Oxford, 1991); K. Mizoguchi, 'A Historiography of a Linear Barrow Cemetery: Structurationist's Point of View', *Archaeological Review from Cambridge* 11 (1992), 39–49; K. Mizoguchi, 'Time in the Reproduction of Mortuary Practices', *World Archaeology* 25 (1993), 223–35.

[63] J. Last, 'Books of Life: Biography and Memory in a Bronze Age Barrow', *Oxford Journal of Archaeology* 17 (1998), 43–54.

[64] H. Härke, 'Material Culture as Myth: Weapons in Anglo-Saxon Graves', in *Burial and Society: The Chronological and Social Analysis of Archaeological Burial Data*, ed. C. Kjeld Jensen and K. Høilund-Nielsen (Aarhus, 1997), pp. 119–27.

[65] L. Hedeager, 'The Creation of Germanic Identity: A European Origin-Myth', in *Frontières d'Empire: Nature et signification des frontières romaines*, ed. P. Brun, S. van der Leeuw and C. R. Whittaker (Nemours, 1992); L. Hedeager, 'Cosmological Endurance: Pagan Identities in Early Christian Europe', *European Journal of Archaeology* 1 (1998), 382–96; K. Høilund-Nielsen, 'Style II and the Anglo-Saxon Elite', *Anglo-Saxon Studies in Archaeology and History* 10 (1999), 185–202.

Increasingly, work has been done on the 'cultural biographies' of objects, to discuss how the use and exchange of objects gives them identities and associations with particular memories. With regard to Stonehenge, for example, Bradley has recently argued that the use of Bluestones was intended to evoke cultural associations with the Preseli mountains of south-west Wales (perhaps in the form of an origin myth).[66] The idea is that the materials contributed something to their new context that evoked associations with their place of origin. John Chapman has recently highlighted the importance of the circulation and fragmentation of both objects and human remains in the process of constructing social relations and concepts of time in past societies.[67] Following from the work of Eves and other anthropologists (see above), Andrew Jones and Yannis Hamilakis have also discussed the importance of the production, circulation, consumption and disposal of food and drink during mortuary rites, for memory work using case studies from Neolithic Orkney and Bronze Age Crete. They suggest that distinctive links between the senses, embodiment and the mortuary events are created by the consumption – in other words, the incorporation – of food and drink at the tomb.[68]

SUTTON HOO RECONSIDERED

Some of these approaches are already in place in early medieval archaeology, while others have yet to be investigated for the period. Yet, in combination, these perspectives provide new trajectories of research into the burial sites of early medieval Britain. In the remainder of this essay, I wish to pursue some of these theoretical ideas through a single case study: the high status barrow burials of the seventh century, focusing upon the familiar and renowned burial site of Sutton Hoo in Suffolk (Figure 1). The cemetery has been the focus of a series of excavations, most famously in 1938–9, but also in 1967–71 and most recently between 1983 and 1993. Full descriptions of the site and its historical context can be found elsewhere, and many of the observations and interpretations made in this study rely on the excellent reports and synthetic

[66] R. Bradley, *An Archaeology of Natural Places* (London, 2000), pp. 81–96; P. Geary, 'Sacred Commodities: The Circulation of Medieval Relics', in *The Social Life of Things: Commodities in Cultural Perspective*, ed. P. Geary (Cambridge, 1986), pp. 169–94; J. Moreland, 'The World(s) of the Cross', *World Archaeology* 31 (1999), 194–213.

[67] J. Chapman, 'The Living, the Dead and the Ancestors: Time, Life Cycles and the Mortuary Domain in Later European Prehistory', in *Ritual and Remembrance: Response to Death in Human Societies*, ed. J. Davies (Sheffield, 1994), pp. 40–85.

[68] A. Jones 'The World on a Plate: Ceramics, Food Technology and Cosmology in Neolithic Orkney', *World Archaeology* 31 (1999), 55–77; Y. Hamilakis, 'Eating the Dead: Mortuary Feasting and the Politics of Memory in the Aegean Bronze Age Societies', in *Cemetery and Society in the Aegean Bronze Age*, ed. K. Branigan (Sheffield, 1998), pp. 115–32.

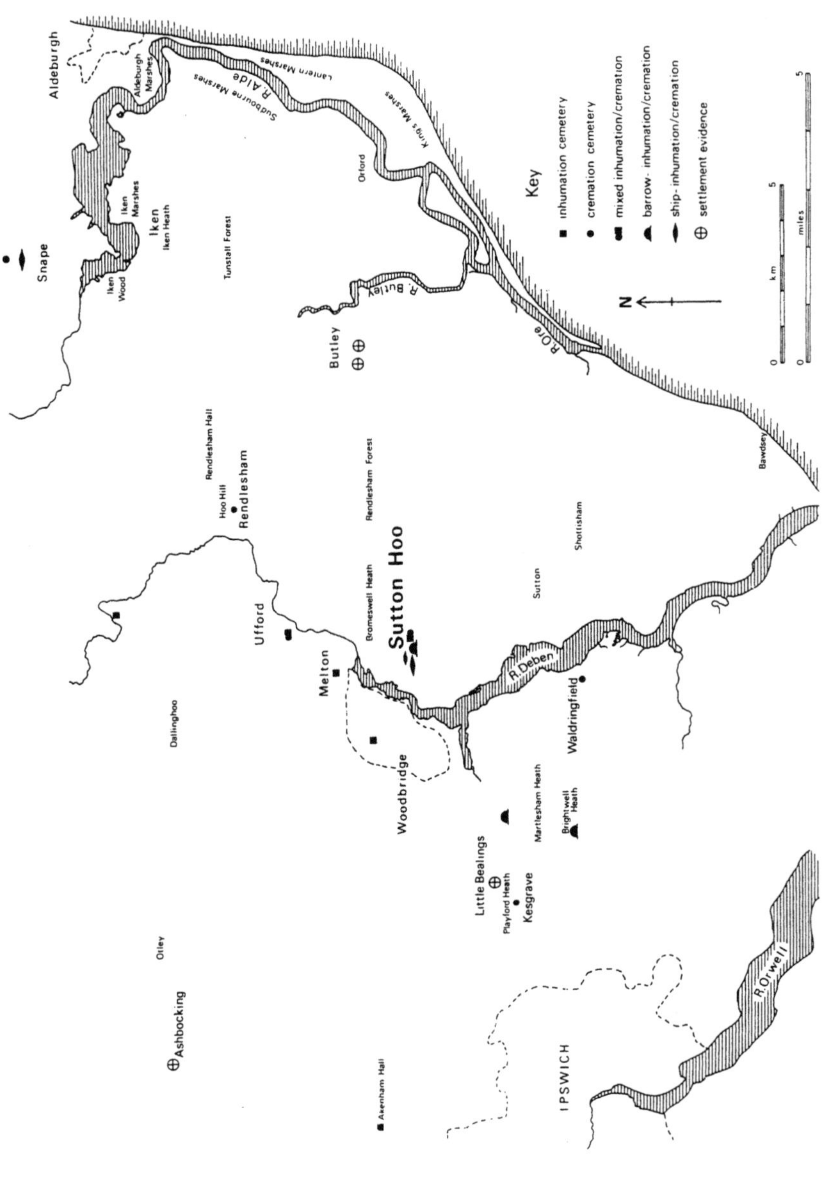

Figure 1 The location of the Sutton Hoo cemetery. After *Sutton Hoo Research Committee Bulletins 1983–1993*, ed. M. Carver (Woodbridge, 1993), Bulletin 1, p. 10.

accounts of Rupert Bruce-Mitford and Martin Carver.[69] The last ten years have seen a renewed interest in this site, with many articles focusing upon the objects and the mortuary rites. The aim of this essay is to attempt to add to the continuing debate concerning the interpretation of the site, in the hope that the theoretical perspectives outlined above will allow some different approaches to familiar evidence. Inevitably, all observations are provisional upon the eagerly awaited publication of the Sutton Hoo Research Trust's excavations by Martin Carver and Madeleine Hummler.[70]

The Sutton Hoo cemetery

The discovery at Sutton Hoo in 1939 of the most wealthy Anglo-Saxon grave ever found (the ship burial in mound 1) has until recently overshadowed the rest of the sizeable barrow cemetery. The supreme wealth discovered within the burial chamber in mound 1 consisted of gold jewellery, silver and bronze vessels, weaponry and other 'regalia' thought to have been buried with an inhumed, presumably male, individual whose remains were dissolved in the acidic soil.[71] However, there are many other barrows that form the cemetery: at least nineteen in total. Of the twelve that have been excavated, all have been dated to the late sixth and early seventh centuries AD. Beneath other barrows were remains of further inhumation graves: another male was buried beneath a ship under mound 2; a much robbed and disturbed female burial was excavated beneath mound 14; and a male weapon burial was interred next to his horse beneath mound 17. These graves display a range of high-status gold, silver and ironwork, even though none proved as opulent as mound 1. The evidence is fragile and partial, given that many of the mounds were disturbed in earlier centuries by grave robbers and the human remains were destroyed by the acidity of the heathland soils. Despite this, these wealthy graves loom large in accounts of the site.

However, Martin Carver has noted that cremation rites were actually more common at Sutton Hoo than inhumation graves. Cremated remains sometimes also contained fragments of wealthy grave goods that had accompanied the deceased on the pyre. The remains were partially retrieved and placed within bronze containers (mounds 4, 5, 6, 7, 18), but in one case the ashes were placed on an oak tray (mound 3) (Figure 2). The relative

[69] R. Bruce-Mitford, *The Sutton Hoo Ship-Burial I: Excavations, Background, the Ship, Dating and Inventory* (London, 1975); R. Bruce-Mitford. *The Sutton Hoo Ship-Burial II: Arms, Armour and Regalia* (London, 1978); *The Sutton Hoo Ship-Burial III*, ed. R. Bruce-Mitford and A. C. Evans (London, 1983); *The Age of Sutton Hoo*, ed. M. Carver (Woodbridge, 1992); *Sutton Hoo Research Committee Bulletins 1983–1993*, ed. M. Carver (Woodbridge, 1993); M. Carver, *Sutton Hoo: Burial Ground of Kings?* (London, 1998).

[70] M. Carver and M. Hummler, *Sutton Hoo: The Early Medieval Cemetery and its Context* (forthcoming).

[71] Carver, *Sutton Hoo*, pp. 144–53.

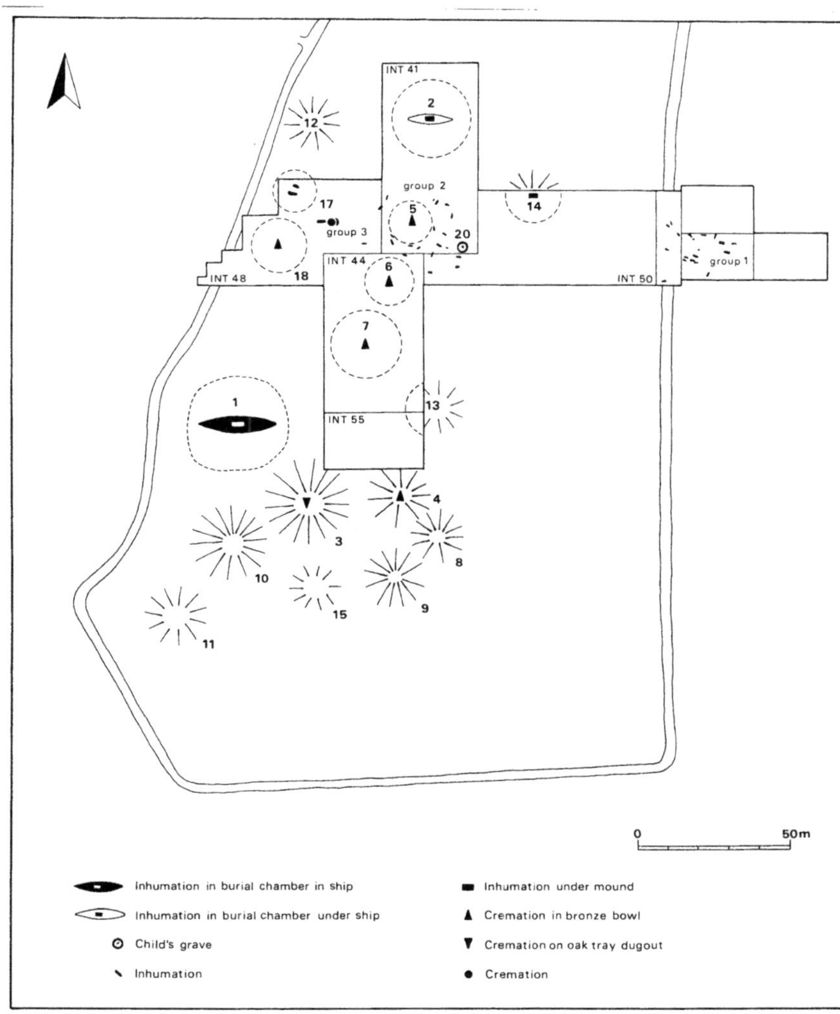

Figure 2 Plan of the barrow cemetery at Sutton Hoo. After *Sutton Hoo Research Committee Bulletins*, ed. Carver, Bulletin 8, p. 6.

paucity of objects in these graves cannot be taken at face value as evidence that the cremation graves represent less wealthy or lower status individuals. The cremation process will have destroyed many of the items placed on the pyre. Indeed, the spectacle of cremation may have made the funeral equally opulent and splendid. This is supported by the fact that the wealth of objects with the mound 3 cremation are exceptional and outstanding for

Anglo-Saxon cremation burials, reflecting a burial rite reserved for the highest echelons of a changing society.

Furthermore, themes run through both inhumation and cremation rites. Animal sacrifice, particularly the slaughter of horses, is found in both inhumation and cremation graves; and, once buried, the remains of both inhumation and cremation graves were covered with comparable mounds, suggesting a similar status for all those buried within the cemetery.[72] This is not to say that the distinction between cremation and inhumation was not socially or ideologically significant, only that there is no reason to assume a status difference between those buried in each manner.

The archaeological interpretation of the Sutton Hoo sites has focused on a number of themes. In many ways, the wealth and scale of the burial rites are unique in early seventh century England, yet there are comparable 'princely' graves found throughout many parts of the Anglo-Saxon kingdoms from Taplow and Caenby to Coombe and Asthall.[73] They are often regarded as exclusive cemeteries, markers of territory containing symbols of power, and hence evidence of the increasing social complexity of early Anglo-Saxon society.[74] They may reflect the emergence of kingship and the territorialization of kingdoms. They can also be seen as a statement of ideological affiliation to Scandinavia and paganism and a resistance to Kentish and Merovingian Christianity.[75] Martin Carver has developed upon these ideas, discussing the nature of the complex rituals at these sites as analogous to theatrical performances and poetic texts, with potentially many shifting meanings for both participants and onlookers.[76] Elsewhere, this author has argued that their precise location in the landscape served to articulate these distinctive ritual performances.[77] As enduring markers in the landscape, it is easy to think of the barrow burials of the seventh century as a prominent

[72] J. Bond, 'Burnt Offerings: Animal Bone in Anglo-Saxon Cremations', *World Archaeology* 28 (1996), 76–88; Carver, *Sutton Hoo*, pp. 107–36.

[73] M. Welch, *Anglo-Saxon England* (London, 1992); L. Webster, 'Death's Diplomacy: Sutton Hoo in the Light of Other Male Princely Burials', in *Sutton Hoo: Fifty Years After*, ed. R. Farrell and C. Neuman de Vegvar (Oxford, OH, 1992), pp. 38–46.

[74] C. J. Arnold, *An Archaeology of the early Anglo-Saxon Kingdoms* (London, 1998); Carver, *Sutton Hoo*; C. J. Scull, 'Before Sutton Hoo: Structures of Power and Society in Early East Anglia', in *Age of Sutton Hoo*, ed. Carver, pp. 3–24; Webster, 'Death's Diplomacy'.

[75] M. Carver, 'Cemetery and Society at Sutton Hoo: Five Awkward Questions and Four Contradictory Answers', in *Spaces of the Living and the Dead: An Archaeological Dialogue*, ed. C. E. Karkov, K. M. Wickham-Crowly and B. K. Young (Oxford, 1999), pp. 1–14; R. Van de Noort, 'The Context of Early Medieval Barrows in Western Europe', *Antiquity* 67 (1992), 66–73.

[76] Carver, *Sutton Hoo*; M. Carver, 'Burial as Poetry: The Context of Treasure in Anglo-Saxon Graves', in *Treasure in the Medieval West*, ed. E. M. Tyler (Yorke, 2000), pp. 25–48.

[77] H. Williams, 'Placing the Dead: Investigating the Location of Wealthy Barrow Burials in Seventh Century England', in *Grave Matters: Eight Studies of First Millennium AD Burials in Crimea, England and Southern Scandinavia*, ed. M. Rundkvist, British Archaeological Reports, International Series 781 (Oxford, 1999), pp. 57–86.

display of the dead, their social and political status and an attempt to emphasize links between the land and kingship.

So far, however, there has been little consideration of the ways in which death and remembrance were articulated through the burial rites. Through the remainder of this paper it is argued that we can identify in the archaeological evidence from Sutton Hoo a complex series of relationships between the mortuary practices and the promotion of a distinctive form of social memory. It is the contention of the present discussion that these observations may hold important clues for developing our appreciation of the social and ideological importance of early medieval cemeteries.

The landscape context

Let us start with the setting of the graves (Figure 1). They are placed on the crest of a terrace overlooking the River Deben, a tidal estuary linked to the North Sea.[78] In comparison with barrow locations on downland in southern England or the Peak District, the situation may appear unimpressive, but within its local topographic context, the cemetery might be described as 'prominent'. While past vegetation patterns are difficult to ascertain fully (as are the precise height and appearance of the mounds when newly built), it is likely that views were possible across the Deben valley; equally, the barrows would have been sky-lined from many parts of the environs.[79] Such a location might be deemed ubiquitous for Anglo-Saxon burial sites, but this is not so. It can be demonstrated that other cemeteries have different, less commanding locations. Furthermore, the size of the barrows at Sutton Hoo would have created a very different visual impression from most other communal cemeteries of the fifth to seventh centuries. Elsewhere, I have argued that such situations may serve to make the dead visible within the landscape and may relate to a belief in their active supervision of the affairs of the living.[80] After all, it is a common conception of the later Scandinavian literature that the dead in part resided in their tombs.[81] Whatever the exact motivation, barrows were meant to be seen from a distance in specific ways. There may also have been a desire to locate barrows with particular kinds of views. In this sense, barrows were places with distinctive ways of seeing the landscape. Such locations would encourage a particular mnemonic relationship between vision and place, since each visit entailed experiencing unique prospects of the landscape.[82]

[78] Bruce-Mitford, *The Sutton Hoo Ship-Burial I.*

[79] Bruce-Mitford, *The Sutton Hoo Ship-Burial I*; Carver, *Sutton Hoo*; Webster, 'Death's Diplomacy', pp. 77–8.

[80] Williams, 'Placing the Dead'.

[81] H. R. Davidson (Ellis), *The Road to Hel* (Cambridge, 1943).

[82] For an anthropological example of the importance of views in ancestral locations, see M. Bloch, 'People into Places: Zafimaniry Concepts of Clarity', in *The Anthropology of Landscape*, ed. Hirsch and O'Hanlon, pp. 63–77.

Although set back from the ridge, it is possible that the cemetery was visible from the Deben to the west, and this particular relationship with a tidal estuary as a physical boundary and a route of movement should be emphasized.[83] It appears as if the barrows would have been visible to travellers moving along the Deben valley, by ship or land. Seagoing ships such as those found under mounds 1 and 2 could have travelled from as far afield as Kent, Francia, Frisia and Scandinavia and could reach upstream along the Deben at least as far as Sutton Hoo en route to Rendlesham.[84] It was here that written sources inform us there was a royal estate centre, and possibly a pagan temple, in the seventh century.[85] Alternately, Sutton Hoo could have been a destination in itself, given the hint of a settlement nearby on the hilltop and a pagan shrine suggested by place names.[86] The burial site may have been much larger than the area covered by the remaining mounds, and earlier Anglo-Saxon graves have recently been found to the north of the site. There may be an additional factor to consider why the barrows were placed exactly where they were. The barrows seem to be situated east of the river at the only point where the main course of the Deben meanders westwards. In other words, for travellers heading upstream from the North Sea towards Rendlesham, Sutton Hoo would have to be passed. At the point of approach, the Deben swings sharply to flow westwards. Travellers moving upstream by the river would move eastwards at this point, so that the barrows would command the skyline directly ahead. This is the only point along its course where the river flows from east to west and consequently may be a deliberate consideration in the siting of the mounds.[87] Moving downstream away from Rendlesham, the barrows would be visible for many kilometres on approach; but because of the lie of the land, the cemetery would disappear from view downstream from this point. So, moving downstream would mean moving away from the barrows until they were out of sight. The location not only 'overlooked' the river and surrounding country, but may have set up certain recognized spatial orientations in relation to existing routes of movement. The mnemonics of this relationship may be seen in the character of the barrows themselves, not only being sky-lined, but also containing ships that must at one time have traversed the same route. It is interesting that the two ship burials known

[83] Bruce-Mitford, *The Sutton Hoo Ship-Burial I*, pp. 1–4.

[84] Bruce-Mitford, *The Sutton Hoo Ship-Burial I*, pp. 24, 30; M. Carver, 'Pre-Viking Traffic in the North Sea', in *Maritime Celts, Frisians and Saxons*, ed. S. MacGrail (London, 1990); Carver, *Sutton Hoo*. See also P. Anderson, 'Aspects of Site Topography and Boat Morphology in the Inhumation Boat Graves of Vendel Period Sweden', in *Vendel Period Studies*, ed. J. P. Lamm and H.-Å. Nordström (Stockholm, 1984), pp. 31–8.

[85] R. Bruce-Mitford, 'Saxon Rendlesham', *Proceedings of the Suffolk Institute of Archaeology and Natural History* 24 (1949), 228–51.

[86] Carver, *Sutton Hoo*, p. 61; P. Warner, 'Documentary Sources', in *Sutton Hoo Research Committee Bulletins 1983–1993*, ed. Carver, Bulletin 2, pp. 7–8.

[87] Bruce-Mitford, *The Sutton Hoo Ship-Burial I*, Plate B.

from Sutton Hoo were both orientated west–east, the orientation of their final approach to the cemetery when hauled up the steep slope from the river.[88]

There has been considerable debate about the precise location of the boundary of the East Anglian kingdom and its relationship with Sutton Hoo.[89] We do not know whether Sutton Hoo lay close to the peripheries of the kingdom or, alternately, whether the barrow cemetery is situated at a local centre, given hints of a high-status settlement on the same ridge and a nearby pagan temple site.[90] Despite this uncertainty, the relationship discussed above between the barrows and the Deben as a route of movement between the sea and a royal vill inevitably means that the barrows marked movement into, and out of, the East Anglian kingdom. They enforced a symbolic threshold on an international stage.

It appears that the barrows were situated to overlook, and to be seen from, routes of movement through the landscape. Their precise situation may have been chosen to enhance this experience and encourage remembrance for those moving to and from major centres of royal authority and power. In this way, landscape and memory were interlinked through the placing of the graves, both in relation to routes of movement and as symbolic thresholds between kingdoms.[91]

Mound building and social space

I have argued that the subtle topography of south-east Suffolk was being employed to project the remembrance of the Sutton Hoo dead onto the wider landscape. It may also be possible to see the working of social memory in a number of ways in the use of cemetery space and monumentality. Recent discoveries have included sixth-century cremation and inhumation burials north of the Sutton Hoo cemetery.[92] This suggests that the Sutton Hoo cemetery was not an exclusive elite burial site, but formed part of an older communal site that may have stretched along the ridge for over 500m to the north of the visible mounds. Even so, it appears that the southernmost extent of the cemetery was not interspersed with many other lower-status graves (Figure 2). This may represent an exclusive burial area within the larger communal burial site. In this exclusive burial zone, we might imagine that burials were less frequent and that their large mounds took up more space

88 Bruce-Mitford, *The Sutton Hoo Ship-Burial I*, p. 22.

89 Bruce-Mitford, *The Sutton Hoo Ship-Burial I*, p. 30; M. Parker Pearson, R. Van De Noort and A. Woolf, 'Three Men and a Boat: Sutton Hoo and the East Anglian Kingdom', *Anglo-Saxon England* 22 (1994), 27–50.

90 Carver, *Sutton Hoo*, p. 61; Warner, 'Documentary Sources'.

91 Williams, 'Placing the Dead'.

92 S. Newman, pers. comm.

when compared with contemporary communal cemeteries.[93] Furthermore, unlike many Anglo-Saxon burial sites where there are multiple graves, with one exception, the barrows at Sutton Hoo seem to have covered the remains of single persons, sometimes with sacrificed animals.[94] There is no evidence of the successive use of barrows for secondary interments apart from a small satellite barrow south-east of mound 5, and two groups of later execution graves (see below). Hence, each barrow represented a single funeral: any subsequent rituals and ceremonies at the burial site, together with the addition of new barrows, would take place in recognition of the presence of the existing barrows and the discrete events they commemorated.

Barrow construction has been regarded in terms of social status and ideological affiliation. But first and foremost, it served to commemorate the funeral and influenced memory in many ways. Observing the barrow would have evoked not only the identity of the individual inhabiting it, but the range of participants and procedures leading to its creation. We know little of the appearance and speed of the mounds' degradation, or whether each barrow was elaborated with posts or other structures. It certainly has been suggested that the mast of the ship in mound 2 could have been in place rising out of the barrow. In what better way would the remaining structure evoke the funeral? There are three factors of special note. First, the substantial frequency of barrow construction contrasts with that found in contemporary burial sites elsewhere and would create a unique impression. Where found elsewhere, such barrows do not seem to occur in cemeteries outside Kent, Sussex and Surrey. Secondly, although there are variations in the size of the barrows, compared with other sites there is a general impression of uniformity in the scale of mound building, reflecting equivalent amounts of labour. Both factors would influence memory, emphasizing not only the importance of the individual funeral and the identity of the deceased but also the relationships between old and new tombs. Thirdly, an inevitable consequence of building such large monuments over single graves was that each burial was much more spatially discrete than usually is the case in fifth- to seventh-century burial sites. This created a different use of space for those visiting and experiencing the site from other contemporary communal cemeteries. These observations may appear obvious, yet they may have had a profound and distinctive effect upon the way that social memory was articulated through the mortuary rites.

We can pursue these themes further by looking at the precise positioning of mounds in relation to each other.[95] The cemetery broadly follows the ridge

[93] If the cemetery was used for only twenty years, on average only one burial would take place each year. In contrast, most communal Anglo-Saxon cemeteries on average may have received somewhere in the region of between two and five burials per year.

[94] The cremation grave beneath mound 4 contained the remains of a man and a woman: Carver, *Sutton Hoo*, p. 110.

[95] Carver, *Sutton Hoo*.

in a crude linear fashion, although no precise order is evident. In the southern half of the cemetery the mounds appear clustered in no clear order. This need not mean that the location of each barrow was not significant, but as Martin Carver's tentative chronological phasing of the site suggests, the cemetery may have developed in a complex way from a number of separate foci. However, in the northern half of the cemetery there seems to be logic in the placing of barrows into two parallel lines following the contours of the ridge (12, 17 and 18, and 5, 6, 7 and possibly 2). From the Deben to the west, and from the land to the east, this linear arrangement would allow social relationships between the barrows, and the memories of each funeral, to be observed and recognized. Elaborate spatial distinctions and associations seem to be at work in the precise positioning of barrows, the exact details of which will remain obscure until a more detailed chronology of the site is forthcoming. In other words, memories of the dead were not simply articulated by raising a barrow, but through the precise location of each new mound in relation to earlier graves.

Finally, we come to the topic of monument reuse. Other high-status barrows of the early seventh century appear to deliberately appropriate and transform ancient monuments. For example, at Taplow the barrow sits within an Iron Age hillfort, and at Snape it focuses upon at least one Bronze Age barrow.[96] There have been hints that a similar practice of monument reuse applies to Sutton Hoo. It has been suggested firstly that some of the barrows may have had an earlier Bronze Age phase, and secondly that Iron Age field systems may still have been visible and influenced the precise location of some mounds.[97] The evidence remains ambiguous, but a reference to the ancient past may be found in the barrows themselves.

I have argued that in a number of ways, the scale and form of barrows served to create mnemonic links between the monuments; but another scale of temporality may be evoked. The size of the barrows means that they are equivalent to the larger Bronze Age monuments found throughout the early medieval landscapes of eastern England such as nearby at Brightwell and Martlesham Heaths where another Anglo-Saxon barrow burial was found in close association with substantial prehistoric barrow groups.[98] Could this allusion have been intentional? If this was the case, then regardless of the presence or absence of an ancient monument on the Sutton Hoo site, building a new barrow may have served to manipulate conceptions of social memory

[96] Excavations in the summer of 1998 revealed evidence of a multivallette hillfort encompassing Taplow Court and the Anglo-Saxon barrow. Pers. ob. For Snape, see W. Filmer-Sankey, 'Snape Anglo-Saxon Cemetery: The Current State of Knowledge', in *Age of Sutton Hoo*, ed. Carver, pp. 39–52.

[97] A. Ellison, 'The Potential for Prehistoric Research', in *Sutton Hoo Research Committee Bulletins 1983–1993*, ed. Carver, Bulletin 4, pp. 39–40; Carver, *Sutton Hoo*, p. 165.

[98] R. Gilyard-Beer, 'The Devil's Ring, Brightwell Heath', *Proceedings of the Suffolk Institute of Archaeology and History* 34 (1979), 247–78.

through the 'antiquation' of the funerary rites.[99] So, barrow building not only created fame and remembrance for the deceased and his/her kin, it also reproduced relationships with earlier funerals through its form and scale, and simultaneously may have alluded to the remembrance of an ancestral or mythical past.

So far we have discussed the landscape setting, spatial organization and monumentality of the Sutton Hoo site. It has been suggested that these combined to create a distinctive experience of the funerals and subsequent engagements with the cemetery that would have manipulated and projected memory work in particular ways. A complex mixture of memories and temporalities seems to be created through the mortuary rites, engaging memories for those travelling at a distance as well as those visiting the site for ceremonies, rituals and funerals. For each of the elements discussed above, Sutton Hoo would stand as relatively unique in comparison with other, contemporary and earlier Anglo-Saxon burial practices. However, this is not the whole story. We also need to consider the way the body was treated, and the employment of material resources within the graves in terms of social memory.

Transforming the body: technologies and artefacts

At Sutton Hoo, we have evidence of two contemporary strategies for transforming the body: cremation and inhumation. Both, in different ways, involved an ideology of transformation. As mentioned above, cremation is the most common burial rite; as also discussed above, transformation by fire in many societies has a profound influence upon how death and the dead are perceived and remembered within a social and cosmological context. Cremation is usually misunderstood in archaeological circles as a destructive process, but in many societies it is deemed more as a transformation that contrives to create something new. Themes of fertility, regeneration, incorporation and embodiment are widespread in post-cremation rites in many societies. The funeral seeks to destroy and rebuild the deceased's body into a new physical and spiritual form.[100] At Sutton Hoo, great attention was clearly paid to the retrieval and collection of the ashes from the pyre and their transportation to the burial site. These rites can be regarded as serving to reconstitute memories of the dead into a new identity as well as incorporating them physically into a new locale, and socially into a wider community history. In short, cremation at Sutton Hoo was a process of memory work and creation.

[99] Williams, 'Monuments and the Past'.

[100] H. Williams, 'Remains of Pagan Saxondom? The Study of Anglo-Saxon Cremation Rites', in *Death and Burial in Early Medieval Britain*, ed. S. Lucy and A. Reynolds (forthcoming). See also Kan, *Symbolic Immortality*.

These themes also apply to the inhumation graves, although in this case there is the emphasis towards the retention of bodily integrity until after burial. Even so, the aim was not to portray the dead as they were in life, but to re-create the dead into a new image and body in the grave. In cases both of inhumation and of cremation, the identities of the deceased are re-placed and re-membered beneath sizeable barrows. The distinction between the two procedures is important, since this may have been another means of creating unique and distinctive memories while recalling elements of funerals that went before. The use of two rites would in itself create an interplay of similarities and differences between each grave. Just as with the raising of a mound, the mode of interment would have created remembrances of, or comparisons with, earlier funerals, or created contrasts between them.

Let us now look in more detail at the objects that were associated with these technologies of the body. Inevitably, because of its nature, we know most about the way objects were used in the burial rite from the excavated inhumation graves, chiefly mounds 1, 2 and 17. What were the specific roles of material culture in this process? The disposal of such a quantity of wealth in itself was a dramatic statement that would have been remembered, and the cremation, procession and/or grave-side rituals would have been impressive spectacles within which such acts of conspicuous consumption would find a place for both inhumation and cremation. In addition, in association with the body, these objects created an 'image of the dead'. For both mound 1 and mound 17, we can chart the complex sequence through which the objects were placed in the grave. The ways in which the dead were represented were slowly built up through a series of distinct social practices. With such assemblages we often only see the final product and forget that memory was constituted through the process of the assemblages' creation. In the case of mound 17 (Figure 3), the spears were followed by a shield, bucket and two cauldrons, a kit-bag, a bronze bowl and a harness. Only then was the coffin lowered in, containing the young male with a sword and purse. Finally, a comb was dropped into the grave.[101] This complex set of rites involves both display and concealment (see also Figure 4). The body was contained within a coffin from an earlier stage of rites, and items within would have been hidden. There were also acts of destruction and sacrifice, since a horse was killed and buried in a separate pit next to the grave. Each act would build and interact with each other to create the image and fame of the grave.

Both the spectacle and display, and the association with the body to form an image of death, served to link objects with social memory; but the real significance of these acts came from their immediate concealment. In other words, all these acts and objects created memories, and displayed together they painted a picture of the dead in the grave. But this image of the deceased

[101] Carver, *Sutton Hoo*, p. 112.

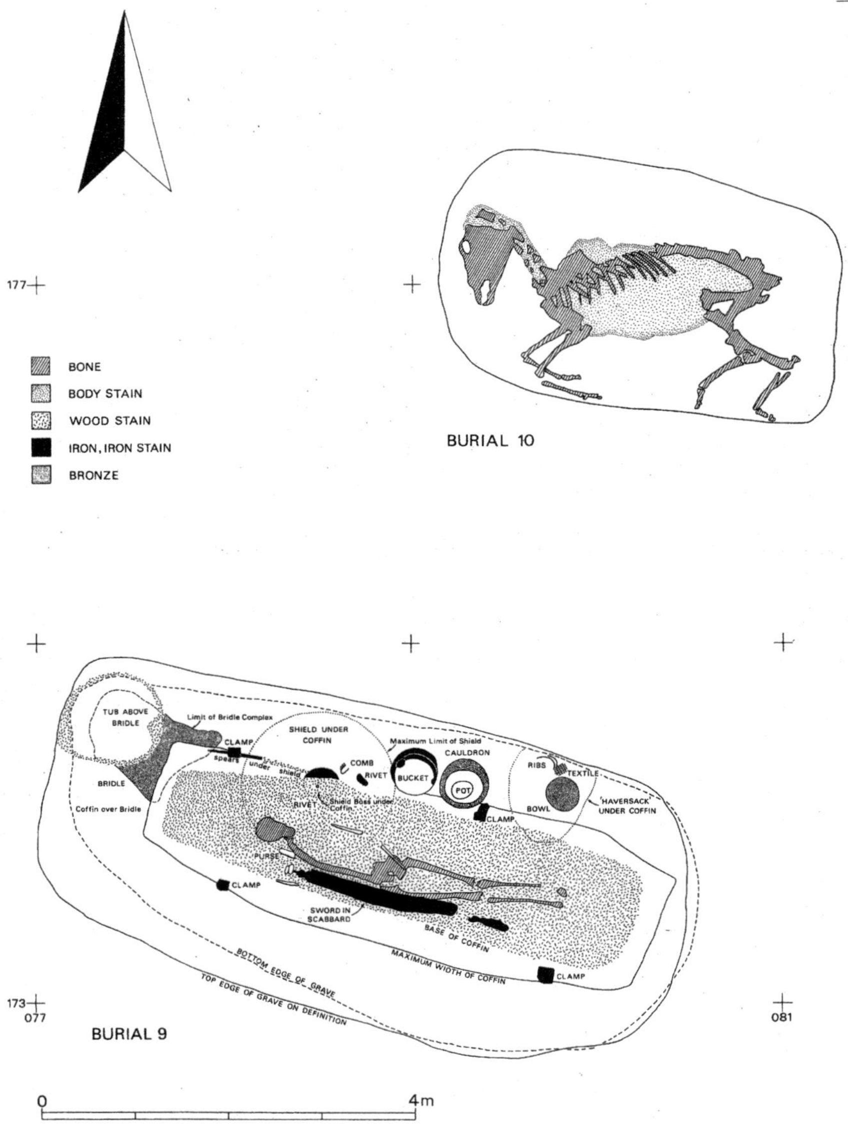

Figure 3 Plan of the male adult inhumation and horse burial beneath mound 17 at Sutton Hoo. After *Sutton Hoo Research Committee Bulletins*, ed. Carver, Bulletin 8, p. 8.

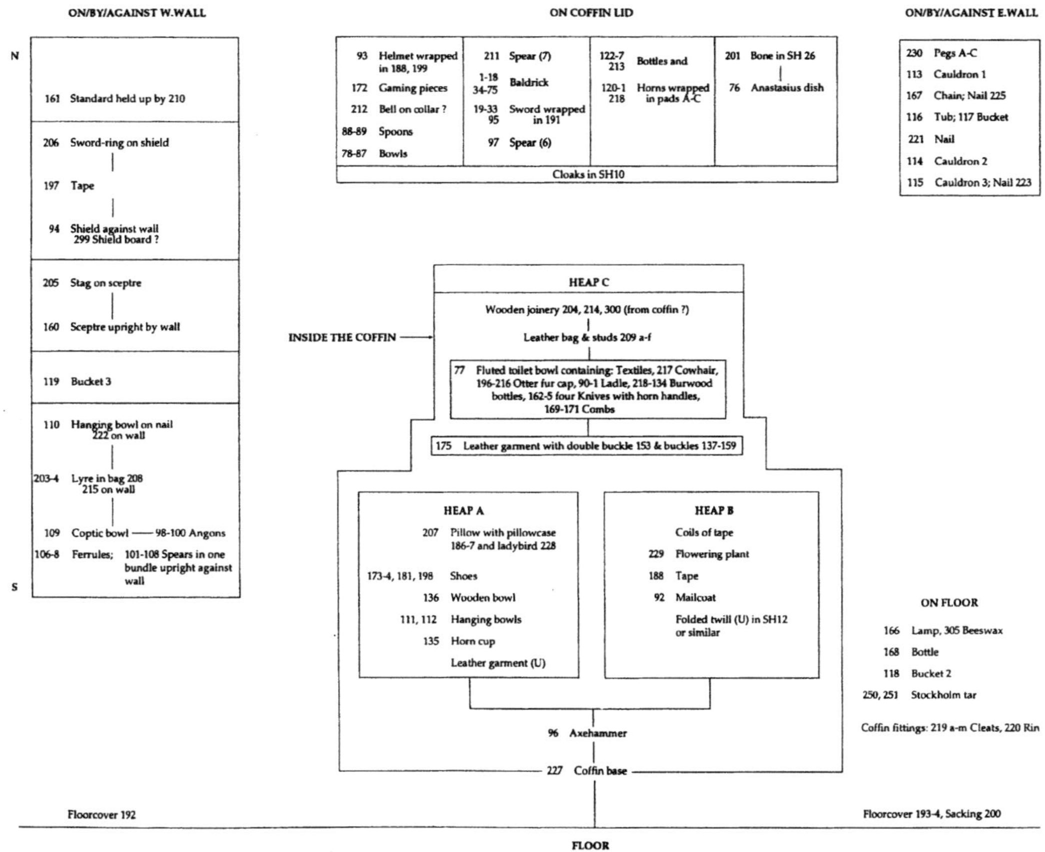

Figure 4 The sequence of objects placed in the chamber of the ship buried in mound 1 at Sutton Hoo. After M. Carver, *Sutton Hoo: Burial Ground of Kings?* (London, 1998), p. 124. Reproduced by permission of British Museum Press.

was significant only because it did not endure. Both the objects and the body were then concealed by earth, so that memories created dramatically through complex sequences of ritual acts were immediately taken out of view. As Küchler notes (see above), the display, followed by the hiding or destruction of objects, can serve as a powerful mnemonic device, translating objects to an existence that can only be recalled through association with the places where they occurred. The only other means was for the evocation of the fame of earlier graves through repeating ritual actions at subsequent funerals. But even this concealment was a structured event, since different objects would have gone out of view at different times. For instance, in the mound 1 ship, certain items (such as the gold coins in the purse) would have been concealed before their inclusion in the grave deposits (Figure 4). Items within the coffin including hanging bowls and the clothing of the deceased would have disappeared from view at an early stage, covered by a sequence of objects placed on the coffin lid, such as the helmet and the Anastasius dish.[102] The objects placed around the wall, such as the cauldrons to the east of the coffin and the sceptre, Coptic bowls and other 'regalia' to the west, would disappear from view with the closing of the burial chamber within the ship, probably while still lit by a ceramic lamp (Figure 4). Finally, and slowly, the ship itself, the container of all these items with their separate memories and associations, would disappear from view. These acts of closure could have been elongated over several hours, several days, or perhaps longer and with each, the 'fame' of the grave would increase as less remained visible.

I have discussed the wealth of objects, their use to create an image of the dead, and their deposition in the mound as elements that served in the workings of memory. Also to be considered are the cultural biographies of the objects themselves and the nature of the image created for the dead through technological processes and the deposition of material culture. It can be argued that social memory was being worked through the different antiquities and provenances of objects used in the funerary rites. Many of the items interred with the dead are likely to be of local production and used by the deceased or kin during their lifetimes. However, we also have a range of more exotic and older equipment, particularly in the mound 1 ship burial but also in cremation mound 3. These include objects that reflect (either through their provenance or by their stylistic affiliations) long-distance contacts with the Merovingian, Mediterranean, 'Celtic' and Scandinavian worlds (Figure 5). Let us take some examples from mound 1. The ship from mound 1 had a working life, with evidence of repair, and many of the artefacts were of some antiquity when buried.[103] For instance, the hanging bowls show signs of repair and the Mediterranean silverware could have

[102] Carver, *Sutton Hoo*, p. 124.
[103] Bruce-Mitford, *The Sutton Hoo Ship-Burial I*, pp. 350, 412–13.

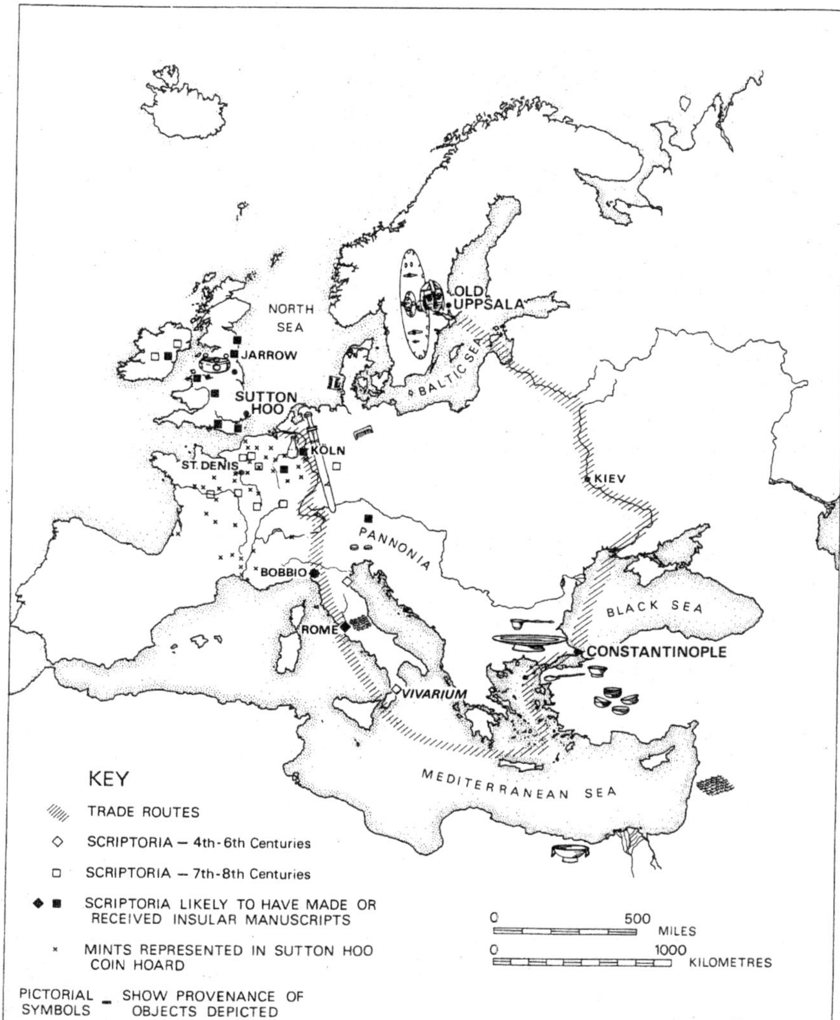

Figure 5 The origins and affiliations of objects from mound 1 at Sutton Hoo. After *Sutton Hoo Research Committee Bulletins*, ed. Carver, Bulletin 4, p. 48.

been up to a century old when disposed of.[104] The hanging bowls incorporated technologies that derive from western Britain or Ireland.[105] The Merovingian coins, each from a separate mint, would (whatever the precise significance) evoke relations with Merovingian Gaul, as might many of the Mediterranean artefacts.[106] The bone casket from mound 3 also suggests connections with Francia or Italy.[107] Other items of gold and weaponry are probably locally made, but bear strong technological and stylistic affiliations with central Sweden.[108] The shield and helmet may be of central Swedish manufacture or followed closely on Swedish designs, and the gold and garnet sword pommel is made of purer gold and may be an import from Scandinavia.[109]

These objects can be seen to reflect the gift-giving practices and long distance cultural interactions between the early medieval elites in north-west Europe.[110] Yet these objects brought to the funeral their own temporalities and mnemonics depending on to whom they belonged, how they were used, the memories of where they originated, and the social relationships they symbolized. Alternatively, in the case of locally made products, they reflected the memories and temporal associations that the elites wished to create for themselves. In this way, the fame of the deceased was created by the objects buried with him.

Other objects from Sutton Hoo brought together powerful associations with the social and cultural activities of elite groups. The deposition of ships and the sacrifice of horses and other beasts may evoke metaphors of mobility, journeying and transition in death,[111] while the weaponry symbolizes the

[104] *The Sutton Hoo Ship-Burial III*, ed. Bruce-Mitford and Evans; R. Bruce-Mitford, 'The Sutton Hoo Ship-Burial: Some Foreign Connections', *Settimane di studio* 32 (1986), 144–210 (p. 157).

[105] M. Ryan, 'The Sutton Hoo Ship Burial and Ireland: Some Celtic Perspectives', in *Sutton Hoo*, ed. Farrell and Neuman de Vegvar, pp. 83–95.

[106] Bruce-Mitford, *The Sutton Hoo Ship-Burial I*, p. 578; D. Whitehouse, 'The Mediterranean Perspective', in *Sutton Hoo*, ed. Farrell and Neuman de Vegvar, pp. 117–28.

[107] Bruce-Mitford, *The Sutton Hoo Ship-Burial I*, pp. 105, 147.

[108] Bruce-Mitford, *The Sutton Hoo Ship-Burial I*, pp. 167–95; Bruce-Mitford, *The Sutton Hoo Ship-Burial II*, pp. 90–9, 224.

[109] Bruce-Mitford, *The Sutton Hoo Ship-Burial II*, pp. 188, 196.

[110] The exact character of exchange is one that is much debated. One possibility often discussed is that we are seeing evidence of retinues. See K. Hoilund-Nielsen, 'Retainers of the Scandinavian Kings: An Alternative Interpretation of Salin's Style II (Sixth-Seventh Centuries AD)', *Journal of European Archaeology* 5 (1997), 151–69.

[111] For grave goods as metaphors, see A. Andrén, 'Doors to Other Worlds: Scandinavian Death Rituals in Gotlandic Perspectives', *Journal of European Archaeology* 1 (1993), 33–55. For cremation and animal sacrifice, see Bond, 'Burnt Offerings'; P. Crabtree, 'The Symbolic Role of Animals in Anglo-Saxon England: Evidence from Burials and Cremations', in *The Symbolic Role of Animals in Archaeology*, ed. K. Ryan and P. Crabtree (Philadelphia, 1995), pp. 20–6; J. Richards, 'An Archaeology of Anglo-Saxon England', in *After Empire: Towards an Ethnology of Europe's Barbarians*, ed. G. Ausenda (Woodbridge, 1995), pp. 51–55; H. Williams, 'An Ideology of Transformation? Cremation and Animal

military strength and ideals of the individuals interred. Of particular interest, the bronze bowls used as containers for cremated remains may allude to other social practices: namely, feasting and drinking. In other contexts, these vessels are found with foodstuffs within them.[112] The presence of drinking equipment and evidence of food offerings are present throughout the cemetery and are often regarded as reflecting elite lifestyles and ideals of the afterlife.[113] However, their particular mnemonic role is not discussed. The association with cremation is especially telling, for the bones of the dead are substituted for the drink and food. Consequently, the use of such vessels to contain the dead may represent some form of metaphor between the consuming and incorporating aspects of feasting, and the transformation and reconstitution of the dead through the cremation process. Be that as it may, the mortuary context evokes elements of an idealized lifestyle during life, as well as long-distance contacts and interactions.

Rituals of memory

The significance of the burial rites and grave goods at Sutton Hoo has already been discussed in terms of origin myth and ideological affiliations. Bruce-Mitford, following Chaney, emphasizes that much of the mound 1 regalia may allude to concepts of sacred kingship and ancestral links with Woden.[114] More recently, Kelley Wickham-Crowley and Stephen Glosecki have discussed the metamorphic and possibly even shamanistic qualities of the metalwork that may pertain to Woden,[115] and hence could have served to evoke associations between the individual buried and a glorious past and divine descent. These arguments need not relate to the mound 1 assemblage in isolation, but to the wider mnemonics and meanings of the cremation rite and the barrow cemetery as a whole. While recognizing the Mediterranean, Frankish and Kentish influences, Martin Carver has stressed the Scandinavian and pagan affiliations of the burials, suggesting from the combination of ship burial, cremation, furnished burial and the stylistic elements of the

Sacrifice in Early Anglo-Saxon England', in *The Archaeology of Shamanism*, ed. N. Price (forthcoming).

[112] J. Musty, 'The Excavation of Two Barrows, One of Saxon Date, at Ford, Laverstock, near Salisbury, Wiltshire', *Antiquaries Journal* 49 (1969), 98–117 (pp. 109–10); H. Geake, *The Use of Grave-Goods in Conversion Period England, c. 600–c. 850* (Oxford, 1997).

[113] C. Neuman De Vegvar, 'The Sutton Hoo Horns as Regalia', and Webster, 'Death's Diplomacy', in *Sutton Hoo*, ed. Farrell and Neuman de Vegvar, pp. 63–74 and 75–81 (p. 79).

[114] Bruce-Mitford, *The Sutton Hoo Ship-Burial II*, p. 371; W. Chaney *The Cult of Kingship in Anglo-Saxon England* (Manchester, 1970), pp. 121–55.

[115] S. Glosecki, *Shamanism and Old English Poetry* (London, 1989); K. Wickham-Crowley, 'The Birds on the Sutton Hoo Instrument', in *Sutton Hoo*, ed. Farrell and Neuman de Vegvar, pp. 43–62.

metalwork from the graves that they represent an ideological resistance to Christian and Merovingian influence.[116] Although he does not go so far as to suggest that a Scandinavian origin myth is being promoted at Sutton Hoo, the adoption of Style II may reflect Hedeager's arguments that it shows affiliations and origin myths pertaining to Scandinavia.[117]

Ian Wood and others have questioned the predominantly Scandinavian associations, arguing for the importance of the Kentish and Frankish connections; this in turn has been used to question the identity of the individual in mound 1 as a son of Raedwald, king of East Anglia, or a member of an East Saxon dynasty.[118] Meanwhile, William Filmer-Sankey argued that the mound 1 individual was dressed and provisioned as a Roman emperor. There are objects from the Mediterranean, namely the Byzantine silverware (two spoons, ten silver bowls, a fluted bowl with a female head, the Coptic bowl and the Anastasius dish); and there are objects of Roman inspiration such as the shoulder-clasps and sceptre.[119] Filmer-Sankey sees this as an aspiration of status, rather than the deliberate manipulation of fame and memory.

To my mind, all these arguments form part of the picture; yet discussions often present a polarity in early medieval cultural affiliations between the Germanic and pagan on the one hand, and the Roman and the Christian on the other. The burial rites may not reflect a pagan ideology of affiliation with Scandinavian, Merovingian or Kentish hegemony, or an image of the Roman emperor, although these elements are present within mound 1 and throughout the cemetery of Sutton Hoo as a whole. There are two points to make. First, the locally produced items that make up the bulk of all the grave assemblages are under-emphasized. It is increasingly clear that many of the elements of the burial rite cannot be as strongly linked to Scandinavia as Carver has suggested. For instance, cremations in bronze bowls of seventh-century date are now revealed as additions to existing communal cremation cemeteries of the fifth and sixth century throughout eastern England, as well

[116] M. Carver, 'Kingship and Material Culture in Early Anglo-Saxon East Anglia', in *The Origins of the Anglo-Saxon Kingdoms*, ed. S. Bassett (Leicester, 1989), pp. 141–58; M. Carver, 'Boat-burial in Britain: Ancient Custom or Political Signal?' in *The Ship as Symbol in Prehistoric and Medieval Scandinavia*, ed. O. Crumlin-Pedersen and B. Munch Thye (Copenhagen, 1995), pp. 110–124 (p. 122); M. Carver, 'Ideology and Allegiance in East Anglia', in *Sutton Hoo*, ed. Farrell and Neuman de Vegvar, pp. 173–82; Carver, *Sutton Hoo*, p. 136; Carver, 'Cemetery and Society', p. 10.

[117] L. Hedeager, 'Cosmological Endurance: Pagan Identities in Early Christian Europe', *European Journal of Archaeology* 1 (1998), 382–96; Hoilund-Nielsen, 'Retainers of the Scandinavian Kings'.

[118] I. Wood, 'The Franks and Sutton Hoo', in *People and Places in Northern Europe 500–1600*, ed. I. N. Wood and N. Lund (Woodbridge, 1991), pp. 1–14; I. Wood, 'Frankish Hegemony in England', in *Age of Sutton Hoo*, ed. Carver, pp. 235–41; Webster, 'Death's Diplomacy'; Parker Pearson, Van de Noort and Woolf, 'Three Men and a Boat'.

[119] W. Filmer-Sankey, 'The "Roman Emperor" in the Sutton Hoo Ship Burial', *Journal of the British Archaeological Association* 149 (1996), 1–9.

as in high-status cremation graves as far afield as Kent and Oxfordshire.[120] The burial rite at Sutton Hoo has a local and regional context, as well as displaying long distance contacts. Secondly, the range of long distance contacts reflected in the mound 1 assemblage bears testimony to a wide range of socio-political contacts and exchanges; it need not reflect any single affiliation with any one region. An understanding of social memory may resolve these contradictory interpretations. The antiquity of both the objects and their stylistic allusions does seem significant. The aim seems not to be to assert one regional or ideological allegiance, but to draw them together in a complex act of *bricolage* – making do with available resources and symbols.[121] The dead were being transformed and created in an ancestral image drawing elements from the Antique, Scandinavian, Celtic and Anglo-Saxon worlds and interpreted within a *local* ritual milieu. It may be the case that the diverse origins were as important in creating a memory for the dead as was the raising of an enduring mound over the grave assemblage. The antique and exotic objects may have served to situate the dead in the past, perhaps akin to dressing them as ancestors. This perspective suggests that objects brought to the funeral, displayed with the dead and finally covered with them, were not simply 'possessions' or 'gifts'; they were mnemonic devices for transforming and manipulating the remembrance of their dead in relation to origin myths, histories and genealogies.

The memory of violence

There is one final consideration to be made of the site and its use. So far I have discussed the memories constructed during the period of the barrow burials, but there is a change in the cemetery's use through the Anglo-Saxon period, with evidence of execution burials in the eastern cemetery and around mound 5 (Figure 6). A wide diversity of postures, orientations and mutilations are present in the graves. In the Eastern Cemetery were 23 graves, including decapitated burials, prone burials with hands behind the back and ankles tied, burials superimposed over each other, an individual with a posture suggesting a broken neck, one in kneeling position with the top of the head missing, and so on. Around mound 5 there are 16 graves including one burial with a possible rope around the neck suggesting a victim of the gallows, decapitated individuals, prone burials, and two females placed face-down over a male burial. The earliest burial west of mound 5 may have taken place soon after the construction of the barrow, but other execution graves in both groups may have continued from the late seventh

[120] T. M. Dickinson and G. Speake, 'The Seventh-Century Cremation Burial in Asthall Barrow, Oxfordshire: A Reassessment', in *Age of Sutton Hoo*, ed. Carver, pp. 95–130 (pp. 128–30).

[121] See C. Tilley, *Material Culture as Text* (London, 1991).

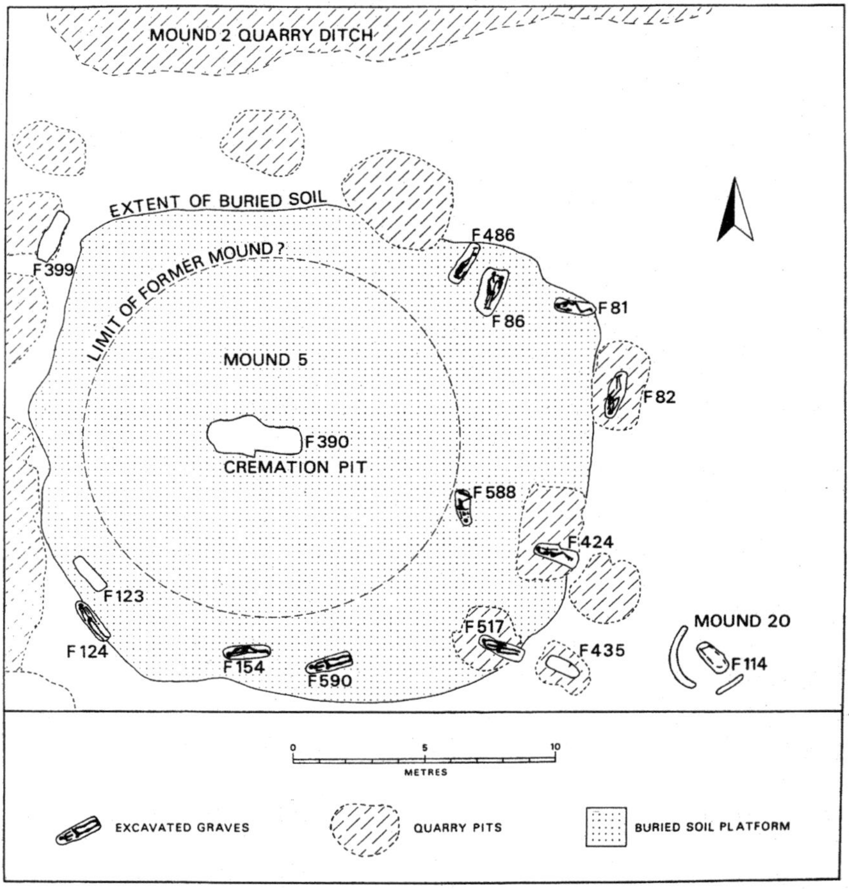

Figure 6 Execution graves around mound 5 at Sutton Hoo. After *Sutton Hoo Research Committee Bulletins*, ed. Carver, Bulletin 8, p. 10.

to the eleventh centuries AD (Figure 6).[122] As Martin Carver and Andrew Reynolds have argued, these ritual killings are as much a statement of kingship and authority as the barrows, even if they show a different category of the dead (political or ideological deviants) employed to assert royal power.[123]

122 Carver, *Sutton Hoo*.

123 A. Reynolds, 'Anglo-Saxon Human Sacrifice at Cuddesdon and Sutton Hoo?', *Papers from the Institute of Archaeology* 7 (1996); A. Reynolds, 'The Definition and Ideology of Anglo-Saxon Execution Sites and Cemeteries', in *Death and Burial in Medieval Europe*, ed. G. de Boe and F. Verhaeghe (Stockholm, 1993), pp. 33–41.

However, for our purpose, this 'afterlife' of the barrows demonstrates a distinctive manner by which social memory was projected and transformed. Carver notes that the earliest execution victim, burial 53, lies within the quarry ditch west of mound 5, the only barrow containing a cremated young individual who had also died a violent death (the bones revealed blade injuries to the skull).[124] Why was mound 5 chosen for the burial of executed individuals, and perhaps used as the killing place? Carver's explanation seems feasible: that the memory of the violent death of the occupant survived beyond the Christianization of the East Anglian royal family.[125] When the site became used for execution of those offending royal authority, it provided an appropriate narrative to evoke at the execution.

We cannot know the exact motivations behind this choice. Reynolds has argued that Christian imagery was an important part of execution cemeteries, both in their location on prehistoric barrows and in the manner of death by hanging. The barrow may have evoked associations with the hill of Golgotha and the death of Christ, and the burial of victims in groups of three may have emphasized this allusion.[126] These were places that were liminal between heaven and earth. They may alternatively have been regarded as dangerous places of suffering, inhabited by demonic and supernatural forces.[127] Furthermore, these were highly visible places for the corpse to be displayed as a reminder to the living.

At Sutton Hoo, elements of this ideology may have been present, but more distinctive references were being made. The public ceremony of execution seemed to have made use of the fragments of memories of the barrow's inhabitant, evoking them in the placing of the execution site and the burial of each victim. Carver suggests that if the occupant of mound 5 had died in trying to maintain his authority, then execution at that spot may have been pertinent to asserting the authority of the current dynasty.[128] A moral and religious motive may have also been present, relating the ritual of execution directly to the occupant of mound 5. The public killing and burial of a criminal in association with a pagan ancestor who had equally met a grisly demise distant from God could have effectively consigned both the mounds and the criminals to a conceptual region 'outside' the Christian kingdom.

[124] Carver, *Sutton Hoo*, p. 179.

[125] Carver, *Sutton Hoo*, p. 143. This practice can be seen as a different, but in some ways comparable, strategy of remembrance to the use of the early seventh century male barrow burial at Taplow as the focus for a church site, possibly in the late seventh or eighth century. Webster, 'Death's Diplomacy', p. 79; D. Stocker and D. Went, 'The Evidence of a Pre-Viking Church adjacent to the Anglo-Saxon Barrow at Taplow, Bucks.', *Archaeological Journal* 152 (1995), 441–50.

[126] Reynolds, 'Anglo-Saxon Human Sacrifice'; Reynolds, 'Definition and Ideology'.

[127] S. Semple, 'A Fear of the Past: The Place of the Prehistoric Burial Mound in the Ideology of Middle and Later Anglo-Saxon England', *World Archaeology* 30 (1999), 90–109.

[128] Carver, *Sutton Hoo*, p. 143.

This might prove an effective statement for the royal dynasty at once of their links with the pagan past and their distinctions from it. Alternatively, the public execution ceremony, as a commemorative practice, may have served to create a moral contrast between the pagan, heroic and famed violent death of a just and royal ancestor and the humiliating demise of the criminal. The execution might be seen as a symbolic retaliation for the honourable death of the occupant of mound 5. As Maurice Bloch shows, such rituals of 'rebounding violence' are commonplace in societies past and present.[129] Either way, we can see a clear example of how the memories of the mounds were important and actively incorporated into a new Christian context of judicial execution, whose purpose in turn was to serve as a act of remembrance of the moral and physical harm that befalls those that displease royalty and God.

Summary

I have reviewed the many ways in which the body, ritual technologies, objects, monuments, space and place were combined to create and reproduce concepts of memory through the mortuary practices at Sutton Hoo. We have seen how each of these elements may have influenced the reproduction of social memory through the evocation of different temporalities. These included allusions to earlier funerals, to the distant past, to distant places, to contemporary social practices and to diplomatic contacts. Together, these constructed an image and created a range of different, interweaving temporalities, from associations with the distant past to the social practices of the individual in life as well as to aspirations of a future life. Those more familiar with the site will probably be able to elaborate upon these observations in more detail, and the publication of the final report on the most recent excavations may render some of these comments misplaced. Drawing all these elements together, however, it can be argued that, individually and collectively, the burials at Sutton Hoo presented a mythical geography and social memory for those using the cemetery. The transformation of the body, employing wealthy and exotic grave goods with powerful metaphors and associations, the construction of a monumental landmark over individual burials, and their exclusive placing in a commanding location combined both to *incorporate* and to *inscribe* memories and histories.

CONCLUSION: MORTUARY PRACTICES AND WRITTEN HISTORIES AS ALTERNATE STRATEGIES OF REMEMBRANCE

This essay has argued that bodily and commemorative practices in funerary contexts were important strategies of remembrance of histories and mythol-

[129] M. Bloch *Prey into Hunter: The Politics of Religious Experience* (Cambridge, 1992).

ogies in early medieval Britain. I have suggested that archaeological evidence can tell us a great deal about the distinctive ways in which death and memory are intimately linked in the reproduction of social and political structures. The case study of Sutton Hoo is but one context in which these ideas have a value. There is great potential in applying these ideas to a wide range of archaeological phenomena, including the cremation rite in the fifth and sixth centuries AD, the archaeology of the cult of saints, and the mnemonics of early medieval stone sculpture.[130] If so, then the earliest histories of early medieval Britain can be seen in a new light. While they represent the first attempts by literate Christians to record histories that promoted the ideological interests of elite groups, they may have developed and continued alongside comparable strategies evident in high-status mortuary rites.

[130] H. Williams, 'Passing Time and Remembering the Dead in Anglo-Saxon England', unpublished seminar paper, York, October 1999.

4

Lawyers' Time in England in the later Middle Ages

Paul Brand

Mic octo quin tres men cras ani cras mar et oc quin
Hil oc quin cra Purif oc Pas quin tres men habet et quin
Trin oc quin Johan cras hic oc numerando diebus
Nona dies sequitur communiter ordine litis
Quinta dies dotis sequitur quam doceo te

No great aesthetic claims can be made for *Dies communes in banco domini regis versificati*, but then elegance was not its point.[1] These verses, written in around 1300, were practical. They were intended to introduce law students to one important aspect of 'lawyers' time' and to give them a ready way of remembering the things that needed to be remembered about it.

LEGAL TREATMENT OF PRESENT TIME

Terms

Let us start with the 'Mic', 'Hil', 'Pas' and 'Trin' of the verse. During the thirteenth and fourteenth centuries (indeed, through to the nineteenth century) the Common Bench at Westminster, the main royal court for the hearing of civil litigation, heard business on a continuous (or almost continuous), day-by-day basis, not all the year round but during four distinct, but unequal, periods of the year known as 'terms' (the law terms of Michaelmas, Hilary, Easter (*Pascha*) and Trinity). Between each of these terms there fell 'vacations', a term certainly in use by 1279.[2] A separate plea roll was compiled for each of the court's justices to record the business heard during each of these terms and, by the last decade of the thirteenth century, reporters were beginning to

[1] For surviving texts of these verses see BL MS Additional 22174, fol. 218r; Cambridge, MA, Harvard Law School, MS 28, fol. 144r.

[2] There was a break between between the Hilary and Easter terms in the 1279 Kent eyre but by virtue of a proclamation and so it was not accounted as a vacation: 'licet esse vacacio ab uno termino usque ad alium terminum et tamen habetur continuacio et haberi debet immediata ac si non esset vacacio propter communem proclamacionem prius factam'. PRO JUST 1/370, m. 10. The implicit contrast is with the procedure of the Common Bench.

report the cases heard there in termly collections of cases.[3] When the court of King's Bench was reconstituted on a permanent basis *c.* 1234, it too adopted a similar pattern of hearing business in termly sessions (though initially the rolls for several terms were typically bound up together) with terms similar in length to those of the Common Bench.

Michaelmas term was the only term to start on a fixed date each year and also to last for a fixed period of time.[4] It was also the longest of the terms. It invariably began one week after Michaelmas, on or around 6 October. Prior to 1230, Michaelmas term commonly continued past Advent Sunday to end as late as 7 December (in 1200 and 1223) or even 9 December (in 1195). Thereafter it normally finished on about 1 December, around the beginning of Advent,[5] allowing for about eight weeks of continuous business. *Hilary* term began one week after the feast of St Hilary, on 20 January, and seems to have lasted until the beginning of Lent.[6] Its length varied therefore with the date of Lent, giving a minimum of two weeks, more normally three, and sometimes (as in 1200, 1224, 1229, 1243 and 1278) as long as four or even five weeks for that term. *Easter* term was of a fixed length of around five weeks but the date when it began varied with the date of Easter. It always began two weeks and a day after Easter Sunday and it regularly continued past the feast of the Ascension to end one week after Ascensiontide. *Trinity* term was the most variable of the four terms. Its opening date (the Monday after a week after Trinity Sunday) varied, as did Trinity Sunday itself, with the date of Easter; but it normally closed on a fixed day (three weeks after the feast of the Nativity of St John the Baptist, on 14 July). Thus its length, too, varied depending on when Trinity Sunday fell in the year concerned, and was anything between three and five weeks.

The Common Bench was doing business in four separate terms from the time of its separation from the Exchequer in the mid-1190s. Prior to that date, when ordinary litigation heard at Westminster had been heard in the Exchequer, such litigation seems (to judge from the surviving final concords, our best evidence for when such business was heard) normally only to have been done during two separate periods which correspond to what would later be the Michaelmas and Easter terms: indeed, some final concords of this period were specifically described as being made at the Easter or Michaelmas Exchequer. Thus of the four terms later standard, two (Hilary and Trinity)

[3] P. Brand, *Observing and Recording the Medieval Bar and Bench at Work: The Origins of Law Reporting in England* (London, 1999).

[4] The following survey of the lengths of terms is drawn from my own unpublished work on the plea rolls of the thirteenth and early fourteenth centuries.

[5] Advent Sunday is the nearest Sunday to the feast of St Andrew (30 November).

[6] But for evidence that the Common Bench was still sitting after the beginning of Lent in 1292, see *Saunderville v. Saunderville* where a third party was admitted to the defence of lands where waste had allegedly been committed on 25 February: PRO CP 40/92, m. 200. Lent had begun on 23 February.

seem to date from the mid-1190s; the other two (Michaelmas and Easter) are the descendants of the biennial Exchequer sessions that go back at least to the middle years of the reign of Henry II and are probably substantially older than that. Not long after the Common Bench pioneered the use of four terms, the Exchequer too adopted this practice and began to record business (on memoranda rolls and Exchequer plea rolls) as though it too was observing the same four terms.

It is rather more difficult to explain why the Common Bench adopted this particular pattern of terms and vacations. Plucknett's article on 'Legal Chronology' in the *Handbook of Dates* suggests that the rationale of the law terms was that they 'were necessarily situated in those portions of the year which remained available for legal business after certain days and seasons had been excluded for one reason or another'.[7] The 'certain days' which influenced the shape of terms were those major feast days of the Church on which no work could be done (including litigation) and while a court might simply not do business if one occurred by itself during term-time, it was better to suspend sessions altogether if there was a whole cluster of them.[8] There were two different kinds of rationale for avoiding particular seasons.[9] One was that the season was one in which the Church had forbidden the taking of certain kinds of oath. This applied to the period from Advent to one week after the feast of the Epiphany; and to the period from Septuagesima until two weeks after Easter. The second was of a purely secular nature. August and September were ruled out because neither lawyers, jurors or litigants 'could . . . leave their estates during harvest nor during the period of stock-taking, rent-collecting and accounting, which centred on Michaelmas' and, rather anachronistically for terms which could be traced back to the thirteenth century, Plucknett also mentioned the 'fear of plague during the hot season in town'. The ending of Michaelmas term in early December may indeed have been in part determined by the ecclesiastical prohibition on the taking of oaths during Advent, though this was clearly not a total barrier to the hearing of legal business for not every stage in every case needed the taking of an oath and, as has been seen, in the early thirteenth century Michaelmas term quite commonly went on into Advent. Lawyers, justices and litigants did not come for the most part from Westminster or nearby and may have expected to spend Christmas and New Year at home: something for which a Christmas vacation which lasted until mid-January made proper allowance. The exclusion of periods of clusters of major feasts provides an

[7] *Handbook of Dates for Students of English History*, ed. C. R. Cheney, rev. edn (London, 1978), p. 65. In the recent new edition of this publication, *A Handbook of Dates for Students of British History*, ed. C. R. Cheney, rev. M. Jones (Cambridge, 2000), Plucknett's essay has been revised by the present author. Subsequent citations are to Plucknett's essay in the 1978 edition.

[8] *Handbook of Dates*, p. 66.

[9] *Handbook of Dates*, p. 66.

alternative explanation: this excluded not just Christmas Day but also 26, 27, 28 and 29 December and 1 and 6 January. But it remains something of a puzzle why business was not resumed until 20 January. Hilary term, as has been seen, did not end where a strict observance of the Church's rules on oath-taking might suggest (before Septuagesima) but it did not run into Lent and there is no obvious reason, other than the ecclesiastical prohibition, for this to have been the case. It was therefore presumably (albeit in an attenuated form) the ecclesiastical rule barring oath-taking during this season of the year that in fact determined the occurrence of this vacation. The beginning of Easter term was delayed by one week longer than was the case in the Exchequer (as was also true of the start of Michaelmas and Trinity terms) but was otherwise clearly mainly determined by when the Easter festivity was at an end: the one week delay may be some sort of reflection of the precedence given to purely financial matters when both terms had been part of the arrangements of the Exchequer. The date of the end of Easter term is clearly determined by the wish (perhaps the need) to avoid the major ecclesiastical festival of Whitsun (not just Whitsunday itself but also the following three days) as well as Trinity Sunday. Although Plucknett thought the summer vacation was caused by purely secular factors, there may also have been some kind of ecclesiastical factor at work here, too, for there is an unsourced legal text (presumably of Continental origin) in an early fourteenth-century English legal compendium which talks of the whole period from the feast of St Margaret (8 or, more probably, 20 July) until two weeks after Michaelmas as a period for harvesting grain and grape when it was prohibited that 'none swear on the Gospel or hold plea or make summons, so that this time is given to God for prayer and the settlement of disputes and to reconcile those who are in discord and to collect the goods from which the people are to live'.[10]

That there was in fact a degree of flexibility about when legal business could be done, which meant that at least part of the vacations of the main common law courts was considered perfectly suitable for the transaction of business of this kind, is suggested by the medieval history of arrangements for hearing petty assizes and a variety of other business in assize sessions. In 1285 the Statute of Westminster II, c. 30, for the first time fixed three specific periods of the year for the hearing of the petty assizes.[11] Since it was part of a wider rearrangement of the assize system which envisaged the justices of the central courts playing a significant role in the assize circuits, the dates chosen for the most part fell within the vacations of those courts: between the quindene of St John the Baptist (8 July) and the Gules of August (1 August)

[10] BL MS Harley 493B, fol. 114r: 'nul ne jurge sur seintz ewangel ne nul seculer play ne teigne ne somuns ne face, issint qe cel tens seit done a dieu a prier e a peyser conteks e de acorder cels qe sunt en descord e pur quyller les biens dunt le people deit vivre.'

[11] *SR*, I, 85–6.

immediately after the end of Trinity term; between the feast of the Exaltation of the Cross (14 September) and the octaves of Michaelmas (6 October), immediately preceding Michaelmas term; and between the feast of the Epiphany (6 January) and the feast of the Purification (2 February), overlapping with the start of Hilary term. When the system was recast in 1293 with permanent full-time justices, these term limitations were removed,[12] but a further reorganization in 1303 brought back the previous assize terms.[13] From the fourteenth century onwards it came to be the norm for assize sessions (whose justices also acted as justices of gaol delivery and *nisi prius*) to be held in most counties only twice each year. By the sixteenth century these sessions had come to take place during the Lent vacation (after, rather than before, Hilary term) and in the summer vacation (in July and early August) but this was probably a recent development because, prior to the Reformation, the taking of assizes during Lent seems normally to have required the specific permission of the bishop in whose diocese the assize was being heard.

Return days

The plea rolls of the Common Bench and King's Bench record business done in these courts under headings which allocate the individual items or enrolments to a relatively small number of 'days' within each term: in Michaelmas term to the octaves and quindene of Michaelmas (that is, one and two weeks after Michaelmas) and three weeks and a month after that feast, to the morrow of All Souls' and to the morrow, octave and quindene of Michaelmas; in Hilary term to the octaves and quindene of Hilary and three weeks after Hilary (otherwise the morrow of the Purification) and in some (but not all) years to the octaves and quindene of the Purification; in Easter term to the quindene of Easter and three weeks, one month and five weeks after Easter and the morrow of the Ascension; in Trinity term to the octaves of Trinity and in some (but not all) years to the quindene, three weeks and one month after Trinity, and to the morrow, octaves and quindene of St John the Baptist. These are the days mentioned (albeit in abbreviated form) in the *Dies communes in banco domini regis versificati* already quoted. In the earliest surviving rolls a heading giving one of these days marks the transition from one day's business to the next, but within a few years the later and larger rolls provide headings for each membrane ascribing the business to one (or sometimes more than one) of these days.

These headings give the impression that the courts were not hearing business continuously, but sitting at roughly weekly intervals throughout each of the four terms. One gets much the same picture from looking at the

12 *RP*, I, 99.
13 *Calendar of Close Rolls, 1302–1307* (London, 1908), pp. 89–90.

other records of the court: the final concords made in the court are similarly ascribed to one of these limited ranges of days (though only from 1200 onwards) and surviving writs, whether original writs issued by Chancery to initiate litigation in these courts or judicial writs issued by the court itself in connection with subsequent stages of litigation, require action to be taken and the writ itself to be returned with a note of the action taken on one of these same days. The internal evidence of the plea rolls also shows that when the court told litigants or jurors to be at court again on a specific day they were similarly told to be in court on one of these days. But the appearance, as legal historians have long known, is an illusion. These are lawyers' 'days', each of which lasted for up to seven calendar days. They only began on the day named and sometimes not even that, for some return days began on the following day, as with the return days denominated in terms of a certain number of weeks after Easter or Trinity, and the business ascribed in our records to these days may have been done at any time during that seven-day period.

The term used by legal historians for these periods of a week is 'return day'. It is not a term used in the Middle Ages but it is a useful one since it allows us to avoid awkward circumlocutions such as 'the first day of this day' (*primo die hujus diei*) (meaning the first calendar day within the period covered by a particular 'return day'), which we find used in a plea roll enrolment of 1303.[14] The technical term used in the early fourteenth century in connection with 'return days' is 'term day' or 'common term day' or 'common day' (*jour de terme; diem communem termini* or *comune jour de terme; comune jour*). There is, however, a degree of ambiguity about the meaning of these overlapping terms. In some instances it seems clear that, although the reference is primarily to one of these specific days as inserted in a writ, it may indeed apply by implication to the whole of the 'return day' period. As Serjeant Mutford said in a case of 1301 on objecting to the form of a writ of mort d'ancestor summoned before the justices of the Common Bench at York for the octaves of Hilary: 'this writ ought to contain a certain day of the week and not a term day, because the justices [will not await] the fourth day to take it as in other writs.'[15] In other cases, however, the reference is clearly only to the first day of the return day period, as in a case of 1303 where Serjeant Pashley objected to an essoin as it ought should have been cast 'three days before the common day' whereas the essoin had been entered on the morrow of All Souls' 'which is a common day';[16] or even more clearly when Chief Justice Hengham said in a reported case of the

14 PRO CP 40/147, m. 168d.

15 BL MS Add. 5925, fol. 27v.

16 BL MS Add. 5925, fol. 153r: '*Passelegh.* Yl nous semble qe yl est entre en noun deu forme, qe cest essone deyt estre resceu treis jours devaunt le commun jour e cest essone est entre lendemeyn des almes qe est commun jour.'

same year: 'The term day was the Monday because the Saturday before was All Souls' day.'[17]

From early in the thirteenth century, if not before, there were complex rules about when within the 'return day' period (or, as will be seen, in one case, before it) certain procedural steps had to be taken. The essoin or formal excuse of bed-sickness (*de malo lecti*), available only in actions of right, had to be presented by the essoiner three days *before* the *first* day of the 'return day'.[18] This caused problems when that meant that the essoin had to be cast before the court was itself in session. In an early fourteenth-century case, there was a long discussion as to whether such an essoin had validly been cast before Chief Justice Hengham at Rotherham while on his way up to York or whether it should instead have been cast before the constable of York Castle or at the Exchequer.[19] Other essoins had at least in theory to be cast on the first day of the return day,[20] although already by *c.* 1260 it was possible for an essoin to be cast and enrolled on the second day of the return day 'by the grace of the justices' (*ex gracia justiciariorum*) when it had not been enrolled on the first day due to pressure of business,[21] and a case of 1274 alerts us to the possibility that an essoin might be cast as late as the third day of the 'return day' provided this was not challenged by the litigant's opponent.[22] The court would, however, not give judgment on the essoin until the 'fourth day' of the return day. A plaintiff who was suing mesne process against a defendant or seeking a judgment on his default had to appear on each of the first three days of the return day before being able to secure judgment for going on to the next stage of process or for judgment to be awarded in his favour by default on the fourth day: hence the standard 'X offered himself on the fourth day against Y' that begins so many plea roll entries. A defendant seems in most, though not in all, cases to have been safe in delaying to the fourth day before making his appearance. Judgment would not be given against him before then and if he did appear only on the fourth day he would not be required to explain away his absence on the first three days of the 'return day'.[23] But there were some

[17] BL MS Add. 31826, fol. 190v: '*Hengham.* Le jour de terme si fut le lundi qe le samadi devant si fut le jour de Almes et le venerdi jour de Tous Cenz nent pledable qe ne deit estre aconte.'

[18] For a clear example see *Year Books of the Reign of King Edward the First: Years XXXII–XXXIII*, ed. A. J. Horwood (London, 1864), p. 13 (and Latin record at pp. 503–4).

[19] *Year Books of the Reign of King Edward the First: Years XXXII–XXXIII*, ed. Horwood, p. 13 (and Latin record at pp. 503–4).

[20] *Bracton de legibus et consuetudinibus Angliae*, ed. G. E. Woodbine, trans. and rev. S. E. Thorne, 4 vols. (Cambridge, MA, 1968–77) (hereafter *Bracton*), f. 344 (IV, 90).

[21] *Radulphi de Hengham Summae*, ed. W. H. Dunham Jr (Cambridge, 1932), p. 16.

[22] *William Corbet v. prior of Llanthony*: PRO CP 40/5, m. 47d.

[23] This was already the rule in *Tractatus de legibus et consuetudinibus regni Anglie qui Glanvilla vocata*, ed. G. D. G. Hall (London, 1965) (hereafter *Glanvill*), pp. 5–6 (I, 7). See also the 1262 case of *Ranulph son of Adam of Blankney v. Thomas son of Alexander of Scopwick*: PRO KB 26/166, m. 29d.

exceptions to this. In the writ of right, for example, the tenant could lose by default for failure to appear on the first day of the return day after he had put himself on the grand assize or offered battle (and perhaps even after an essoin *de malo lecti* or requesting and obtaining a view of the property in dispute).[24] If the defendant turned up on the first day of the 'return day' but the plaintiff did not and was not essoined, the defendant would be required to make an appearance on the second and third days and would only be able to secure a judgment on the non-suit of the plaintiff (generally, dismissing the case) on the fourth day. But a stricter view was taken of the plaintiff's obligation to appear on the first day of the return day. He could only in general 'save' his default by producing a special excuse for his absence on the first day or by persuading the defendant to waive that default.[25]

It is this which helps to explain the argument we find in a case of 1300 about when exactly a non-suit should be considered as having taken effect in a prior action. A writ had been acquired from Chancery on the third day of the return day on which the plaintiff had been non-suited in a previous related action. The plaintiff argued that by then the previous action was no longer pending (and so his new writ was valid) because he had by then been non-suited because of his absence on the 'first day' of the return day. The defendant argued that at that time the action had still been pending and so this writ should be dismissed, since judgment dismissing the previous case had only been given on the fourth day. The justices agreed that the previous writ was still pending until the fourth day since otherwise the justices would have had no power to give their judgment dismissing the case.[26]

Archbishop Islip's list of 'excluded days' from the middle of the fourteenth century suggests that in practice only six of the seven days included in most 'return days' would have been available to the courts for the transaction of business, since all Sundays were counted within that category.[27] This had certainly not always been the case. Between 1195 and 1200, final concords made in the Common Bench give the actual dates on which they were made (rather than return days) and it is plain from these that during this five-year period the court commonly dealt with at least this kind of business, and probably with other types of business as well, on a Sunday.[28] There is also some evidence from the earlier part of the reign of Henry III that shows the

[24] *Radulphi de Hengham Summae*, pp. 32–3; BL MS Add. 31826, fol. 164v; *Roger de Pengersiek v. John of Leicester and wife Joan*: PRO CP 40/158, m. 48d.

[25] For evidence that if the defendant appeared on the first three days and the plaintiff did not appear until the fourth the plaintiff was non-suited see the 1290 case of *Master Henry Huse v. Peter of Otford*: PRO CP 40/83, m. 85d (and see also PRO CP 40/130, m. 243d).

[26] *Thomas of Helmingham v. abbot of West Dereham*: PRO CP 40/131, m. 243d.

[27] *Handbook of Dates*, p. 65.

[28] These are tabulated in *Pleas before the King and his Justices, 1198–1212, III*, ed. D. M. Stenton, Selden Society 83 (London, 1966), pp. cii–clxv.

court dealing with business on a Sunday.[29] Passing references prove that the justices of the General Eyre were still hearing business on Sundays at the end of the reign of Henry III.[30] There is passing reference in the 1285 Warwickshire eyre to a case being heard on the Sunday after the octaves of the Purification.[31] Early in the fourteenth century the Warwickshire assize justices, William de Mortimer and Roger de Beaufou, were still hearing business on Sundays. Robert of Staverton, the rector of the church of St Nicholas in Warwick, appeared in their court and attempted to stop the justices holding pleas within his parish on a Sunday. Not only did the justices not heed his admonishment: they promptly arrested him but then bailed him for appearance before them at Coventry some months later.[32] The entry recording this is also of interest since it proves what can normally only be suspected: that the dates ascribed to sessions of the assize justices are the dates of the beginning of their sessions and do not mean that they did not continue over a longer period. Here the entry recording the rector's appearance on the Sunday appears on a roll apparently ascribing its business only to the Saturday after the Translation of St Thomas Martyr 1306.

It had, however, clearly become the general practice by the second half of the reign of Edward I for the Common Bench and King's Bench not to sit on Sundays.[33] We learn this both from the Edwardian law-book *Fleta* and from something Chief Justice Hengham said in passing in response to a complaint in 1289;[34] and there are a number of reported cases of the late thirteenth and

[29] In 1229 the surviving plea roll has its final business allocated to the Sunday after the octave of Sts Peter and Paul; in 1232 and 1233 final concords were made on the Sunday after octaves of Hilary; in 1233 final concords were made on the Sunday after Translation of St Thomas the Martyr; in 1238 final concords are dated to the Sunday after Sts Simon and Jude.

[30] PRO JUST 1/736, m. 10 (bottom): memorandum noting that on the Sunday morning after the octaves of Martinmas the sheriff still had 124 writs and that on that day nineteen writs were pleaded. Note also evidence for a wager of law being adjourned to the following Sunday in the 1268 Wiltshire eyre: JUST 1/998A, m. 2d.

[31] When letters of attorney for one allegedly going overseas were produced the defendant said that he had not gone abroad but that on this very day (specified as being the Sunday after the octaves of the Purification) he was at Stratford outside London in Essex and issue joined on this: PRO JUST 1/958, m. 18.

[32] PRO JUST 1/965, m. 19: 'Robertus de Staverton' persona ecclesie sancti Nicholai Warr' venit die dominica in plena curia coram justiciariis hic et ex parte sancte ecclesie inhibuit justiciariis ne aliqua placita tenerent infra parochiam suam per diem dominicam etc. Ideo committatur gaole. Et super hoc Johannes de Broghton' manucepit habendi corpus suum apud Coventr' in crastino Purificacionis Beate Marie etc.'

[33] For land being replevied before the justices of the Common Bench on a Sunday in Easter term 1277 see PRO CP 40/19, m. 75d.

[34] *Fleta II*, ed. H. G. Richardson and G. O. Sayles, Selden Society 72 (London, 1955), c. 35 (p. 137): 'Et sciendum quod ordinatum est quod ipsi justiciarii non teneant placita sua per dies dominicos . . .' Sunday was also described in passing as a *dies non placitabilis* by Chief Justice Hengham when answering a case brought against him in 1290: PRO JUST 1/541B, m. 21d.

early fourteenth century which only make sense on that supposition. Sundays sometimes fell on what would otherwise have been the first day of a return day. From what Hengham says in the a report of a 1303 case it can be deduced that if the first day of a return day fell on a Sunday the Monday was taken as the first day (and presumably subsequent days adjusted accordingly).[35] But there is conflicting evidence from the years around 1300 as to what happened if the Sunday fell on some other day within the initial four-day period of the return day. Enrolments of cases heard in 1290 and 1296 seem to suggest that if a Sunday occurred within this period its place in the sequence of days was simply taken by the following day, so that if the first day of a return day occurred on a Saturday, the second day would be taken to be the Monday, the third the Tuesday and the fourth the Wednesday and that appearances would have to be made on all four days.[36] In a case of 1302 involving one of the justices of the court (plus his wife), however, where the claimants put in appearances on the first day of the return day (which fell on a Saturday) their appearance on the Monday was described as being on the 'third day' and they asked for judgment by default when the tenant's attorney appeared only on the Tuesday, suggesting the alternative procedure of including Sunday in the sequence of days, but not requiring litigants to appear on that day.[37]

Islip's list of 'excluded days' also included a number of other saints' days and festivals that fell within law terms.[38] It is not clear that all of these led to suspension of the courts' business. Our best guide as to the days which were observed by the courts in the later Middle Ages is probably Chief Justice Coke's early seventeenth century list of major feast days when the courts did not sit: All Saints' Day (1 November) and All Souls' Day (2 November) in Michaelmas term; the feast of the Purification (2 February) in Hilary term; Ascensiontide in Easter term; and the feast of the Nativity of St John the Baptist (24 June) in Trinity term.[39] The clearest evidence for the observance of one of these major feasts in the years around 1300 comes from a report of the objections made to the casting of an essoin *de malo lecti* in Michaelmas term 1303. An essoiner said he had come on All Saints' Day 1303 to cast his essoin

[35] In a report of 1303 Chief Justice Hengham notes that when All Souls' is on Saturday the 'term day' (the morrow of All Souls') is the following Monday: BL MS Add. 31826, fol. 190v.

[36] This seems to be the calculation in *master Henry Huse v. Peter of Otford* in Trinity term 1290: PRO CP 40/83, m. 85d. The same method of calculation seems to be applied in a case of 1296 when a woman is said to offer herself on the fourth day of the quindene of Hilary and this is said to be a Tuesday, though the quindene of Hilary fell on the preceding Friday and the Tuesday would have been the fifth day if Sunday had been included in the calculation: CP 40/111, m. 40.

[37] *Peter Mallore and wife Maud v. Roger Quintyn*: PRO CP 40/141, m. 97.

[38] *Handbook of Dates*, pp. 65–6.

[39] E. Coke, *The First Part of the Institutes of the Laws of England*, 5th edn (London, 1656), p. 135a.

not to the court itself but to the 'hostel' of Ralph of Hengham, who promised that he would bear 'record' that it had been cast and have it entered the following day. Had the court been open on All Saints' Day, clearly only the court would have been the appropriate place for doing this, and the essoin would have been entered at once.[40] Lesser feast days which were also in theory 'excepted days' were probably not observed: this was the case with the feasts of St Luke (18 October), Sts Simon and Jude (28 October) and St Andrew (30 November) in Michaelmas term; St Mark (25 April), Sts Philip and James (1 May) and the Invention of the Cross (3 May) during Easter term; or Sts Peter and Paul (29 June) and the Translation of St Thomas the Martyr (7 July) during Trinity term.[41]

Most of the business ascribed to each return day was probably heard within the seven-day period, but even that is not certain. Legislation of 1275 implies that the justices of the Common Bench were sometimes (presumably as a result of the pressure of business) postponing the adjudication of some business from earlier return days in order to deal with business scheduled for the succeeding days. They had to be prohibited from doing this by c. 46 of the Statute of Westminster I.[42] The legislation may have been observed in practice, but there are grounds for scepticism. A number of contemporary law reports show cases being re-heard on the following day or on another day and sometimes on a succession of other days. Some of these days may have fallen within the return day week but it seems unlikely that all did. In 1289, Chief Justice Hengham confirmed that it was not the practice of the courts to enrol adjournments within a term, as though this was perfectly normal and not confined to adjournments from one day to another within a particular return day; and he was referring to an adjournment which had been from the Thursday before All Saints' (27 October) to the Friday after All Saints' (4 November) and thus in practice from one return day period to another.[43]

By the middle of the thirteenth century, if not before, the system of return days that were really weeks formed part of a wider set of arrangements and procedures that appear to have been intended to ensure, as far as possible, a

[40] BL MS Add. 31826, fol. 190v.

[41] For the period between 1195 and 1200 when it is possible to see from final concords when the court was holding sessions on a daily basis, it is possible to see that the Common Bench sat on St Luke's day in 1195, 1196 and 1199; on the feast of Sts Simon and Jude in 1195, 1196, 1197, 1198 and on St Andrew's day in 1199 (in Michaelmas term); that it sat on St Mark's day in 1197 and 1198; on the feast of Sts Philip and James in 1197 and 1198 (this was also the case in 1219); and on the feast of the Invention of the Cross in 1197 (in Easter term); on Sts Peter and Paul in 1196, 1199 and 1200 and on the Translation of St Thomas in 1196 (in Trinity term).

[42] *SR*, I, 38.

[43] PRO JUST 1/541B, m. 21d. The Thursday before All Saints' fell within the return day of a Month after Michaelmas; the Friday after All Saints' within the return day for the morrow of All Souls'.

relatively even distribution of business throughout each of the terms. When litigation came up for a hearing in the court and was to be adjourned for a future further hearing there (either because of the defendant's absence or for the appearance of a jury etc.), the court might have used a set period of time as the period for the adjournment (though there would have had to be some degree of flexibility about this as the Common Bench was in session for less than half the year); it might simply have set the interval for the next hearing of the case by reference to the distance of the county concerned from Westminster; or it might simply have chosen days at random or accepted proffers from litigants for speedier hearings in their cases. It did none of these things. What it did instead was to follow, in general, a system that adjourned all, or at least most, cases from any individual return day to the return day that fell eight return days after it, irrespective of the amount of calendar time that represented. This is the general scheme of adjournments that is mentioned briefly in *Dies communes in banco domini regis versificati* ('Nona dies sequitur communiter ordine litis') and is laid out in detail in the prose *Dies Communes de Banco.* This begins: 'If a writ comes at the octaves of Michaelmas a day will be given to the octave of Hilary' ('Si breve venerit in octabis sancti Michaelis dabitur dies in octabis sancti Hillarii') and then goes through each of the return days in turn giving the corresponding day to which cases from that day would be adjourned. It was copied into many of the private statute books of the later thirteenth century and the fourteenth century.[44] *Dies Communes de Banco* is, however, a schematic account of how the system worked rather than any kind of continuing practical guide, for it only takes inadequate account of the way Hilary and Trinity terms differed in any given year: thus the adjournments recorded on the plea rolls do not always match those given in this work.

It is difficult to believe that this system of regular adjournments can have simply 'emerged'. It looks as though someone or some group, at some stage in the late twelfth or early thirteenth century, must have drawn it up. But if so, we have as yet no idea of who, or exactly when, and there is no trace of any authorizing legislation. What we do have is the legislation of the mid-thirteenth century that introduced the first general amendments to it. Clause 7 of the draft legislation prepared for the Candlemas Parliament of 1259 proposed setting a norm of adjournments to a return day just three (rather than eight) return days distant when a widow was bringing an action for dower and claiming that she possessed none of it (the action of dower *unde nichil habet*).[45] The clause had undergone significant alteration by the time the penultimate French text of the Provisions of Westminster was prepared for the Michaelmas Parliament of 1259.[46] What was now proposed was that

[44] One version of this is printed in *SR*, I, 208.

[45] P. Brand, *The Making of the Common Law* (London, 1992), p. 360.

[46] Brand, *Making of the Common Law*, p. 363.

adjournments be made so as to allow the case to come up for hearing at least four times in a year and never allowing the case to be adjourned beyond the end of the immediately following term. The effect was to lengthen the average interval between hearings in dower cases from a maximum of three return days to a maximum of rather over five return days. As thus amended the legislation passed into law as clause 6 of the Provisions of Westminster of 1259.[47] When the legislation was reissued in 1263 there was a substantive alteration to this clause, increasing the target number of days to a minimum of five, with six specified as the desirable goal. The clause was amended in a very clumsy way. The wording of the original clause was retained and the new sub-clause with the alternative target was simply tacked on at the end, as though there was no contradiction between them. This was re-enacted in 1267 as part of c. 12 of the Statute of Marlborough.[48] The pseudo-statutory *Dies Communes de Dote* reflects this legislation, with adjournments in such pleas being made to a return day four return days distant. Again, this represents an idealized version of the actual practice of the court, rather than a useful practical guide to clerks and justices when making adjournments.[49] It is also to this legislation that our verse referred when it explained that *Quinta dies dotis sequitur*.

During the years prior to 1259, the justices had used their discretion in a fairly systematic way to favour litigants who were bringing actions (whether by the assize of darrein presentment or by the writ of *quare impedit*) to assert a right to present their candidate to a vacant ecclesiastical living, though the shorter adjournments given were not uniform. The need for speed in such cases was obvious. If a church remained vacant for longer than six months, the bishop was entitled under canon law to collate his own candidate to the vacant benefice, and a litigant who was successful after the six months had elapsed would at most recover damages and the right to re-present the rector who had been collated by the bishop. Clause 7 of the draft legislation prepared for the Candlemas Parliament of 1259 proposed the legislative confirmation of the existing practice of giving shorter adjournments in cases about vacant churches and setting a norm of adjournments to a return day three days distant.[50] The clause had undergone significant alteration by the time the penultimate French text of the Provisions of Westminster was prepared for the Michaelmas Parliament of 1259. Clause 7 of this draft envisaged that adjournments of as little as two weeks (normally two return days) might be given provided the county concerned was close but three weeks was retained for more distant counties.[51] As thus amended, the

[47] *Close Rolls 1259–1261* (London, 1932), p. 148.
[48] *SR*, I, 9 (n. 13) and 23.
[49] *SR*, I, 208.
[50] Brand, *Making of the Common Law*, p. 360.
[51] Brand, *Making of the Common Law*, p. 363.

legislation passed into law as clause 7 of the Provisions of Westminster.[52] In 1267 it was re-enacted without alteration as part of c. 12 of the Statute of Marlborough.[53]

Other variations from the norm are to be explained by the continuing discretion enjoyed by the justices in making adjournments. The author of *Fleta,* writing in the early 1290s, believed that the justices were not entitled to show any flexibility about making adjournments[54] and at least two complainants in 1290 alleged that William of Brunton and his colleagues, the justices of the Common Bench, had committed misconduct by making shorter than normal adjournments in litigation before them. However, Brunton's claim that this was a matter within the discretion of the justices and that they were entitled to take account of the proximity of the county concerned was accepted,[55] and this seems to have remained the position throughout the later Middle Ages.

Real time

Real time and real days are not wholly and completely absent from the records of the courts of Common Bench and King's Bench. Sometimes the court and its clerks found it necessary to note the precise calendar day (though given in terms of a saint's day, rather than a particular day of the month) on which something occurred even if the entry itself took its place on a membrane ascribing the enrolments on it to a generic return day or pair of return days. This was for example commonly the case when lawyers who had been convicted of misconduct were sent to gaol under the provisions of the Statute of Westminster I, c. 29, for a year and a day: actually doing 'lawyers' time', as we might say. Thus, for example, when Peter of Luffenham was sent to gaol during Trinity term 1291 for tampering with the court's essoin roll, the record notes that he was sent to gaol on the Wednesday before the feast of St Margaret the Virgin [in 1291] and also that he was released on the Friday before the same festival in the twentieth regnal year [1292].[56] Another kind of situation where the court was careful to note the actual day when something happened was where a third party was admitted to defend a case in place of the defendant named as tenant of the land in the plaintiff's writ when the latter had failed to put in an appearance. This procedure (known as receit)

[52] *Close Rolls 1259–1261,* p. 148.

[53] *SR,* I, 23.

[54] *Fleta II,* ed. Richardson and Sayles, c. 35 (p. 137): 'Et sciendum quod ordinatum est .. quod nulli detur tardius vel maturius quam alii, set omnes dies sunt pariter omnibus communes in hac forma.'

[55] *State Trials of the Reign of Edward the First, 1289–1293,* ed. T. F. Tout and H. Johnstone, Camden Society 3rd s. 9 (London, 1906), pp. 1–5 and 18–23.

[56] PRO CP 40/90, m. 146d. See also *In Re Gerin le Lindraper*: CP 40/91, m. 210; and *In Re Roger Plat*: CP 40/125, m. 236d.

was introduced in 1285 by the Statute of Westminster II, c. 3, for those with reversionary rights in land (where the current tenant of the land was a tenant for life or term of years or a tenant in tail) but abuse of the procedure led to a statutory modification of the procedure in 1292 (the so-called Statute *de Defensione Juris*).[57] This authorized the justices (as from 4 February 1292) to require sureties from any third party wishing to intervene for the value of the lands concerned as from the day he was 'received' to make the defence up until the day final judgment was given in case the judgment given went against him, and thus provided full compensation for the additional delay he had imposed on the plaintiff. This new procedure was already in operation by the end of that same Hilary term, when a third party requested to be admitted to defend a waste case on the Monday after St Matthew's day (25 February 1292) and was admitted though there was trouble about the sufficiency of the sureties he had found.[58] Sometimes litigants or others were required to appear in court on specific calendar days, rather than return days. This seems regularly to have been the case in actions of right if the parties chose to resolve their dispute by judicial combat between their champions. In a case heard in Michaelmas term 1291, for example, in which there were two separate wagers of battle, the champions in one were adjourned to the Wednesday after the octaves of Hilary 1292 (and the eventual concord reached on the battlefield on the Tuesday after the octaves of Trinity 1292) and the champions in the other to the Thursday after the octaves of Hilary 1292.[59]

The day in court

There may even have been something slightly artificial about the lawyer's day. In the fifteenth century, we have Fortescue's word for it that the courts sat to hear business only for three hours each day, from eight o'clock until eleven o'clock (evidently measuring this by clock time and thus for the same length of time whatever the season).[60] Bolland long ago noted an isolated piece of Year-Book evidence that showed that the courts may sometimes have been in session again in the afternoon,[61] but was unable to show that this was ever a regular occurrence. The thirteenth and early fourteenth century practice seems to have been for the court to sit from *hora prima* to *hora nona* of the day (from first light until the day was three-quarters over), which must have meant that the length of the court day varied from time to time of the year and would in the latter part of Trinity term have been significantly

[57] *SR*, I, 73–4 and 110.

[58] *Saunderville v. Saunderville*: PRO CP 40/92, m. 200.

[59] *Nonant v. Stobbehylle et al.*: PRO CP 40/91, m. 45d.

[60] Sir John Fortescue, *De Laudibus Legum Anglie*, ed. S. B. Chrimes (Cambridge, 1942), pp. 128–9.

[61] W. C. Bolland, *A Manual of Year Book Studies* (Cambridge, 1925), p. 98.

longer than it was in the fifteenth century.[62] The court evidently rose in order to take the main meal of the day (*prandium*) and, even if the court did not sit again after that meal was over, there are later thirteenth century references which show that the clerks of the court did resume work then, not in the court itself but in the chief justice's lodging or elsewhere, and that this was the time when they wrote the judicial writs issued by the court.[63] These same entries indicate that it was also the time when professional attorneys active in the court might expect to gain access to the court's records to check the enrolments relating to their clients.

LEGAL TREATMENT OF PAST TIME

So far, this essay has been considering some of the ways in which medieval courts and medieval lawyers handled what was for them present time: the workings and significance of their division of the year into terms and vacations; the importance in the business of the courts and in the records of the courts of the division of terms into notional 'days' whose real duration was not a day, but a week; the duration of the court's actual day. The second half of this study looks at how lawyers and the courts handled what was already for them past time: what concepts they applied to it and what importance they attached to particular ways of specifying it and dividing it up.

Specification of days and times of the day

When litigants brought litigation in medieval royal and other courts alleging a particular act of wrongdoing (a trespass or a wrongful distraint) or the breach of some kind of agreement, the courts required the litigant or, more commonly, their lawyer, to enlarge in their count to the court in various ways on the rather vague allegation or claim made in the original writ initiating the litigation. One of those ways was by giving precision as to *when* the alleged act of wrongdoing or wrongful distraint had taken place or when the alleged agreement had been reached. In most actions it was enough to specify the day and the year when these had taken place: almost invariably in the form not of a numbered day of a specific month,[64] nor of a day numbered in accordance with classical practice (by reference to the ides, nones or kalends), but by reference to a specific saint's day or festival or the morrow or eve or some other day of the week before or after that day or festival and a reference to the

[62] For evidence for these hours of sitting see *Radulphi de Hengham Summae*, p. 32 and *John Skinner v. Alan Pyot* heard in Hilary term 1305: PRO CP 40/154, m. 122d.

[63] See *in Re Robert of Greshope:* PRO CP 40/90, m. 57d (Trinity 1291); *in Re Reginald of Kerneyek*: CP 40/124, m. 22d.

[64] But for a replevin case expressed as a particular day in a particular month: see PRO CP 40/27, m. 137.

current king's regnal year (or more rarely a regnal year of his predecessor) or, where appropriate, just 'this year', meaning this regnal year and not this calendar year. The fact that many saints had more than one feast day provided a hidden pitfall for the unwary. In 1280 a defendant asked for judgment of a plaintiff's count in an action of replevin alleging unjust distraint, seeking to have the action dismissed on the grounds that he had alleged a distraint made on the feast of St Stephen the Martyr but had failed to specify whether he meant the feast in August, the Invention of St Stephen on 3 August, or the feast in December, the straight feast of St Stephen on 26 December.[65] No judgment is recorded on this objection but it is in two other cases of private criminal prosecution (appeal) of the early 1290s where the plaintiffs (appellors) made a similar kind of error. In a case of 1290 the ground for quashing an appeal alleging the homicide of the appellor's father was that the appellor had said that the killing had taken place on the Sunday before the feast of St Michael the Archangel but had failed to specify whether he meant the feast of St Michael at the end of September, Michaelmas proper, or 29 September, or the feast of St Michael *in Monte Tumba*, on 16 October;[66] while in another case of 1291 one of the grounds given for quashing a widow's appeal for the death of her husband was that she had failed to specify whether the feast of St Andrew on which the killing had occurred was the feast of his Passion or his Translation.[67]

It was in the same form of action (the private appeal) alone that the courts expected even greater precision. In such cases, at least from the middle of the thirteenth century onwards, the appellor was required to specify not just the day when the alleged felony had taken place but also the hour of its commission. Here, too, avoidance of ambiguity was essential. In 1284 an appeal was quashed when the felony was alleged to have been committed 'at the hour of vespers' (*ad horam vesperarum*) on the, at first sight, surprising grounds that this was not a specific hour of the day.[68] The reasons for this are, however, explained in a second appeal quashed on similar grounds in 1292, where the defendant objected to the uncertainty of the hour alleged on the grounds that in some parts of the country the bell was rung for vespers immediately after mid-day ('in aliquibus partibus pulsatur ad vesperas statim post horam meridionam') while in others the bell was rung for this purpose only just before sunset ('in aliis partibus parum ante solis occasionem').[69] If the latter was indeed what the appellor meant, ambiguity

[65] PRO, CP 40/34, m. 26d.

[66] PRO CP 40/83, m. 66d.

[67] PRO CP 40/87, m. 44. The feast of his Passion is 30 November. The *Handbook of Dates* does not include any reference to such a second feast of the Translation. Since the widow also specified a Thursday and a day in the current regnal year, it seems fairly clear that she meant the feast of the Passion on 30 November 1290.

[68] PRO CP 40/54, m. 1d.

[69] *William son of Alan of Holland v. Richard Godgomen*: PRO CP 40/96, m. 147.

could be avoided by specifying 'at the hour of vespers, namely between the dog and the wolf' (twilight) as appellors did in 1301 and 1303,[70] or simply 'between the hour of the dog and the wolf' as an appellor did in 1275,[71] or even (for this must mean much the same thing) 'at the hour of sunset' (*ad horam occasus solis; al hure del soleyl rescosant*) as another appellor did in 1297.[72] As these cases might suggest, there seems to have been no generally accepted convention for using one or other of the different ways of referring to specific times of day, for we find both references to the motion of the sun (at or around mid-night;[73] at first light/dawn [*in aurora diei*];[74] between mid-day and the setting of the sun[75] [surprisingly imprecise]) and references to the canonical offices (at the hour of matins;[76] at the hour of prime or the first hour of the day;[77] between prime and terce).[78] Earlier practice does not seem to have required the hour to be specified. The specimen form of appeal included in *Bracton* does include this feature ('at such an hour')[79] but the author later says that it is only sometimes mentioned and is not essential, and even suggests that any variation in this particular part of the appeal between an initial appeal made before the coroners and the appeal made before the justices will not necessarily be fatal.[80]

In other kinds of case, the closest to any kind of precision about the time of day is the allegation that a particular (and allegedly wrongful) act was committed 'at night'. Allegations that a particular distraint had been made at night are very rare: only one in well over two thousand replevin cases heard in the Common Bench during the reign of Edward I.[81] The explanation appears to be relatively simple. As a note from a legal manuscript of the period explains, 'No man can avow a distraint made at night because this sounds in felony.'[82] The purpose of bringing an action of replevin was normally to determine whether or not the reason given for a particular distraint was justified (to test whether a particular service was owed; whether or not a particular man was entitled to common his animals at a particular

[70] *a houre de vespre, cestasaver entre chien e lou*: PRO CP 40/136, m. 190; *infra horam vesperarum inter canem et lupum*: KB 27/172, m. 61.

[71] *Alice daughter of Eve v. Gerard de Furnival*: PRO CP 40/10, m. 24.

[72] PRO CP 40/121, m. 131 in Latin; the French comes from a French report of the same appeal in BL MS Stowe 386, fol. 153v.

[73] PRO KB 26/147B, m. 3d.

[74] PRO CP 40/11, m. 26d.

[75] PRO KB 26/172, m. 10.

[76] PRO JUST 1/1256, m. 64.

[77] PRO CP 40/23, m. 18d.

[78] PRO KB 26/149, m. 24d; KB 26/169, m. 8d.

[79] *Bracton*, II, 388.

[80] *Bracton*, II, 395.

[81] PRO CP 40/153, m. 235d.

[82] BL MS Add. 31826, fol. 235v notes that: 'Nule home peut avouer destresce fete de nuyst pur coe qe ele soune en felonie.'

location; etc.) and thereby to settle the underlying question of entitlement as between the parties. An allegation of distraint 'at night' made this impossible, for it made it impossible for the defendant to admit (or avow) the distraint, since he would thereby be admitting the commission of larceny. No felony was implied by an action allegedly taken at night to knock down a hedge or wall, but the fact that the action was taken at night was legally important and would need to be specifically denied by the defendant if he was to justify his action as taken to preserve an existing right of way. Immediate self-help action taken to preserve a right of way by destroying an obstruction was avowable if done by day, but not if done by night. This is something asserted by counsel, and not denied by other lawyers or by the justices, in a case heard in the 1293 Staffordshire eyre;[83] and it explains the specific denial that similar action had been taken at night that is found in a case of 1303.[84]

In most cases, although it was essential to mention the day when an act or an agreement took place and any count that omitted to specify them could be quashed as defective, little of any importance depended on the particular date mentioned, and it was generally not possible for the defendant to take issue with his opponent by denying only the *day* of the alleged wrongdoing (for this was taken to constitute an admission that an act of wrongdoing had indeed taken place and the act still remained wrongful) or the day the agreement was made (for this in effect conceded the existence of the agreement).[85] But in a few cases the date mentioned was clearly significant. Take the trespass case of 1296 in which Henry of Caldecote alleged that Robert of Hoo had taken goods and chattels comprising wheat, barley, rye, beans, oats, oxen and pork and silver and gold utensils to the value of £60 from him at Onehouse in Suffolk over thirty years previously, on the Sunday after the feast of the Annunciation in 1264. The date was important because it established that the alleged incident had taken place during what was still peace-time. Robert could not deny that there had been an incident around that time when the plaintiff had been robbed of his goods in which he had been involved, but attempted to establish that it had taken place later, during war-time (the enrolment does not give a specific date). He also claimed that he had then still been so young (aged twelve) that he had not known what he

[83] A note from the 1293 Staffordshire eyre that explains that if a hedge is raised where a neighbour has a right of chace to his common and he knocks it down freshly during the day he does no wrong but if he knocks it down at night this is wrongful even if it has been wrongfully raised wrongly: *Year Books of the Reign of King Edward the first: Years XX and XXI*, ed. A. J. Horwood (London, 1866), p. 463.

[84] PRO CP 40/146, m. 41.

[85] Cf. BL MS Add. 31826, fol. 226v: 'Omission fet en conte contant dunt la partie prent sa excepcion parmy quel nul avantage ne ly acrest ne nuist point al demandant: gracia, omission du jour en ple de dette abate le cunte e jatardeys par travers del jor ne doune pas avantage al defendant parmy sun respuns: nam dicendo quod tali die non mutuavit ab eo pecuniam non est accidens exitus. Responseri potest contra istud quod hujusmodi sunt ad curiam et non ad partem et ex quo satisfactum est curie sufficere debet.'

was doing (*mali inscius*) and had simply joined an armed band of men as it passed through Bury St Edmunds while he was still in the abbot's wardship.[86] A series of early fourteenth century trespass cases allege the grazing of the plaintiff's corn by the defendant's animals. All seem to involve land over which the defendant arguably had rights of common, but on which the plaintiff had chosen to grow a crop of grain. Here the date on which the alleged act of grazing had taken place was crucial. If it had been sufficiently late in the year for the corn to have become ripe (*engrane*) then, so the justices seem to have held, the grazing could not under any circumstances be justified; only if it had taken place earlier than this was grazing of the corn legally equivalent to grazing the grass that should have been growing there.[87]

The specific date on which a distraint had been made was also of some significance in replevin cases though it is only regularly recorded as part of the count in such cases from around 1250 onwards. In a number of Edwardian cases defendants were able to secure the dismissal of the case on the grounds that the distraint of which the plaintiff was complaining had only taken place *after* the day the plaintiff had obtained his writ (and thus that the writ was not a proper warrant for determining whether or not they had been unjustly taken and wrongfully detained before the writ's purchase). But the date mentioned as the date when the distraint was made could also be a trap for the unwary defendant. In a case of 1303 a defendant was careful to deny the date of the alleged distraint and to suggest another because the date of the local court judgment on which he relied as authorization for the distraint made was after the date of the alleged distraint he had made.[88]

Time of peace and time of war

So much for references in legal proceedings to actual, specific past calendar time: specific days in specific years and specific hours or times of the day on those specific days. Legal proceedings also sometimes referred to and utilized certain broader categories of past time. One of these broad categories, in this instance a binary category, that was prominent and important in certain particular contexts, was the concept of a 'time of peace', as opposed to a 'time of war'. The need to draw a clear distinction between 'time of peace' and 'time of war' was most urgent in the context of the criminal law. The only known medieval legislation on this subject was drawn up in November 1267, at the same session of parliament at Marlborough that saw the enactment of the better-known Statute of Marlborough, though it does not seem to have

[86] PRO CP 40/122, m. 32. The jury found that the incident had taken place on the day alleged by the plaintiff and that he was already then aged eighteen.

[87] For these cases see BL MS Add. 37657, fol. 174v (and PRO CP 40/141, m. 43); BL MS Add. 31826, fol. 143v and Cambridge University Library, MS Ee.6.18, fols. 57r–v (and PRO CP 40/143, m. 58d); BL MS Add. 31826, fol. 328r.

[88] PRO CP 40/146, m. 9.

been published until the royal justices in eyre began their first post-war sessions in February 1268.[89] It established the general principle that no-one was to suffer loss of life or members or incur permanent imprisonment for offences committed in war-time, whether the offence was prosecuted by individual complaint or appeal or the subject of a presentment, and whether the offence was robbery, homicide, breach of the peace or some other criminal offence. Such offenders were not automatically to be exempt from punishment, however, for the justices were told that they might impose some lesser punishment at their discretion. The justices were also told what was to count as the 'time of war' for the purposes of the legislation. 'Time of war' commenced on 4 April 1264, when Henry III had left Oxford with banners unfurled heading for Northampton, and it continued until after the battle of Evesham, when, on 16 September 1265, the king in the presence of his barons at Winchester had affirmed and proclaimed his peace. However, in addition to this period of war proper, two other periods were also to count as times of war for certain purposes and for certain specific groups of offenders. The first was the period starting on 4 June 1263 and continuing up to 4 April 1264, when some persons (Montfortian adherents) had robbed and killed and imprisoned their opponents, both ecclesiastics and laymen. Such offenders were to enjoy a similar qualified privilege against capital or corporal punishment for their offences provided they were politically motivated in their offences (or, as the legislation puts it, if they were committed *sub spe guerre*); but for ordinary offenders without this motivation the period was to be treated as a 'time of peace' and they were to be subject to the same punishment as normal. The second period was the period after the public proclamation of the king's peace on 16 September 1265. Here the exceptions were rather more limited and the legislation merely confirmed arrangements that had already been made exempting members of various specific groups from the full rigours of the criminal law. Those who had held out in the Isle of Axholme or the Isle of Ely or had been at Chesterfield or Southwark were to enjoy the qualified privilege against criminal liability for their actions granted by the Dictum of Kenilworth or by individual royal grant. Those who had joined the earl of Gloucester for his march on London in early 1267 were to enjoy the terms of the peace between the earl and the king.

The distinction between actions committed in time of peace and those committed in time of war was also important, as has already been seen, in actions of trespass; but the line could be a difficult one to draw and was not a simple matter of chronology. In the Westmorland eyre of 1278–9, Gundred of Stirkland brought an action against Thomas of Musgrave alleging that on

[89] For enrolments of the legislation as sent to the justices in eyre in Yorkshire and Wiltshire on 3 February 1268 see PRO JUST 1/1050, m. 8d; JUST 1/998A, mm. 18, 41d. A copy of the legislation was also sent to the justices of the 1276 London eyre: *Calendar of Close Rolls 1272–9* (London, 1900), p. 333.

27 February 1264, during the period of quasi-peace, Thomas had come to Great Stirkland and assaulted her and carried off her goods and chattels (forty oxen, twenty cows, 200 sheep, 100 pigs, woollen cloths and vessels of gold and silver) and carted her off to King's Meaburn until she had paid a ransom of 120 marks.[90] A contemporary law report of the case tells us that Thomas immediately raised the defence that, if any such action had been committed (which he did not admit), it had been committed in war-time and so he was not obliged to answer for it. Gundred seems to have had two answers to this: that this was not a 'time of war' according to the statute (the report says the Statute of Kenilworth, but the reference is to the legislation of 1267) and that they had not been on opposite sides in the war because she had not been involved in the war and so he should not enjoy this privileged exemption anyway. Thomas answered that this had been a time of war and that she had become involved in the war because she had harboured certain named individuals who were opposed to the king and had attacked royalists and committed arsons and homicides and other criminal acts. Gundred admitted receiving the particular individuals concerned but said that they had come to Stirkland as the executors of her late husband and to administer his estate and without her asking them to: in other words, she had had no choice and had quite properly allowed them entry into her manor (whatever they might subsequently have done). The report suggests that it was only then that Thomas did what one might have expected him to do at the beginning of proceedings: pleaded a quitclaim of all wrongs committed against her from the beginning of the world until that day. The plea roll more plausibly suggests that the concord was only made after the parties appeared in court. In a second trespass case started in the 1278 Hertfordshire eyre but ultimately determined in the Common Bench in 1280, Philip Taylor of London alleged that Peter Picot had taken from him cloth to the value of eighteen marks on 29 April 1267 in London. Peter's defence was that he had been with the earl of Gloucester at that time, in his march on London, and that this was for these purposes to be treated as a time of war; he pleaded Henry III's grant of peace to the earl and his men for any offences they had committed and the earl's letters patent showing that he had then been a member of his household. Philip, however, countered that the terms of the pardon only covered offences committed against those actively involved in the fighting on the king's side and denied he had then been so involved. When a jury confirmed this, Philip was awarded a judgment for the recovery of his chattels plus damages.[91]

Complex legal problems could also arise in connection with property deeds that had been made during war-time. In 1264 Roger de Mortimer

[90] PRO JUST 1/982, m. 2d. For a report of the case (printed from just one of the surviving MSS) see *Brevia Placitata*, ed. G. J. Turner, Selden Society 66 (London, 1951), pp. 210–12.

[91] PRO CP 40/34, m. 53d. For the prior stage see JUST 1/323, m. 4.

took Adam Despenser and Osbert Giffard prisoner at the battle of Northampton and subsequently held them for ransom at his castle of Wigmore. William d'Evreux, who had fought on the king's side at the battle of Lewes but was subsequently assigned to act as one of the keepers of the captive Henry III by Simon de Montfort, then arranged to hand over two of his own manors to Roger (Stoke Lacy and Lawton) as surety for the 1,000 mark ransom of the two men and agreed that Roger should hold them until the ransom was paid. Each of the men handed over one of his Gloucestershire manors (Adam the manor of King's Stanley; Osbert the manor of Pitchcombe) to William as a security that the ransom would be paid. Unfortunately for William's heir, the charter his father made said nothing of the condition or the intention of the parties that the manors be returned to William when the ransom was paid; it took the form of an unconditional grant. William was killed at the battle of Evesham and after his death Adam and Osbert re-entered their lands. William's son, William, brought an assize of mort d'ancestor to recover them but his claim was dismissed on the grounds that his father's interest in the land had originated in a grant made for the purpose of ensuring the payment of a ransom and during war-time and such grants were null and void. From the mid-1280s to the mid-1290s William's son was engaged in litigation against Roger's son, Edmund, and his widow Maud and against a grantee of one of the manors in an attempt to recover them. During the same period William's mother, another Maud, was also engaged in attempting to recover her dower share of these lands against the same Edmund. The litigation is complex, much too complex to attempt to summarize here.[92] But at the core of each of the cases lay a dispute about the validity of William's original grant to Roger de Mortimer in 1264. On the face of it this was a straightforward grant made between two men who were on the same side in the Civil War and could not be impugned on the grounds that it had been made under any coercion exercised by Roger against William (whether imprisonment or some other threat): even if it was made during war-time the grant was perfectly valid. William tried a variety of answers to this, indeed they varied from case to case. He tried to argue that there was legislation invalidating all deeds made during the period between the battles of Lewes and Evesham. He said that it had been decided by the king and his council that charters or other muniments made at that time were of no validity ('per dominum regem et consilium suum decretum est quod carte sive alia monumenta facta eodem tempore nullius sunt momenti'). This was a dangerous line, for there was no such legislation: the Dictum of Kenilworth (to which he was apparently referring) only voids a much more narrowly defined category of charters. His serjeant (Howard) argued in the 1292

[92] The most important stages are recorded in PRO KB 27/33, m. 22d ; CP 40/24, m. 10; CP 40/53, m. 69; CP 40/60, m. 125; CP 40/86, m. 53d; CP 40/92, mm. 110–110d; JUST 1/303, mm. 18, 19, 38d.

Herefordshire eyre that deeds made during the period were invalid because there was at that time no law in the kingdom which could validate them for the king who was in control of the law was then under custody (in prison) and thus the law itself in prison: but this would call into question the validity of all other actions committed under colour of law (such as court judgments) from that period. He even tried to argue that his father and Roger had in effect been on separate sides since his father had been one of the king's custodians and Roger had been one of those trying to free him. William eventually recovered at least one of the manors concerned.

Most of these cases were brought by writs which specifically questioned the validity of the original grant by William to Roger on the grounds that the lands had been granted in war-time and by reason of the ransom of Osbert or Adam. Such actions were quite extraordinary and exceptional in making a specific reference to a 'time of war'. 'Time of peace' was, by contrast, a standard part of the language of a large number of legal cases. Generally, the kind of title made by claimants in suits to assert their right to any particular form of property based that claim on the seisin (the 'rightful' possession) of themselves or one of their ancestors or predecessors at some specific time in the past. The fact that the possession needed to have been 'rightful' might have made it otiose to spell out what that should perhaps always have implied, that the possession which had been enjoyed was during peacetime, and not solely during a time of war, when possession might have been obtained by force. However, both litigants and the royal Chancery under certain specific circumstances went out of their way to ensure that a litigant's claim specifically asserted that fact: as though this might be needed to concentrate the minds of the jury or of the court on that aspect of the claim. It seems to have been the practice (at least from the time of the earliest surviving plea rolls) for all claimants in actions of right, the most solemn and decisive form of property litigation, to assert that the seisin on which they relied, whether their own or that of an ancestor or predecessor, was a seisin enjoyed 'in time of peace'.[93] Although *Bracton*'s specimen count for the writ of right omits this feature, the omission seems to have been accidental for his later discussion of possible omissions from the count assumes that it is present and explains its presence in relation to the civil war of John's reign.[94] It is also probably only accident (or more likely clerical abbreviation) that explains its absence from a number of enrolments of counts in writs of right of the thirteenth century.

A similar feature also appeared in the closely related claimant's count in the various different types of writ of entry.[95] One, but only one, of the petty

[93] For an example from 1196 see *Curia Regis Rolls*, 18 vols. (London, 1923–99), VII, 329. But it does not appear either in either of the counts on actions of right in *Glanvill*: the action of right for land (II, 3 [p. 23]) and the action of right for an advowson (IV, 6 [p. 46]).

[94] *Bracton*, IV, 169, 171.

[95] See, for example, the specimen counts in *Brevia Placitata*, pp. 16, 40, 60, 146, 168, 179–80.

assizes, the assize of darrein presentment, which was about the right to present a candidate to a vacant church, likewise required the jurors to state which patron had last presented a rector to the living now vacant *in time of peace*. Unaccountably, the form of writ given in *Glanvill* omits this particular phrase but a prior reference to the assize in the treatise suggests that by the time the treatise was completed it was already a standard element in the writ formula provided by Chancery;[96] and it is certainly also mentioned in the first surviving writs of darrein presentment which come from 1199.[97] But even where seisin 'in time of peace' was not specifically asserted in the plaintiff's count or in the writ which he had obtained to initiate his litigation, a denial that the seisin relied on had been in time of peace was still a possibility for the defendant. In 1291 Sarah, the wife of Roger fitzPeter, brought an assize of mort d'ancestor against Robert de Vere, earl of Oxford, claiming one third of the Essex manor of Beaumont as the sister and heir of Robert of Creake and on the basis that he had died in seisin of it. The earl's defence denied the implicit claim. Her brother had not died seised of the land in time of peace and (the earl added) the only interest he had ever possessed had been obtained by an intrusion made by force and arms in time of war five days before the battle of Lewes in which he had been killed. It was this issue that went to the assize jury, making them concentrate on the circumstances under which Robert of Creake had gained possession of the property and the legal status of the period when he had done so.[98]

During the reign of a named king

Actions of right and writs of entry also regularly located the seisin on which the claimant was relying by means of another kind of categorization of past time: by referring to the seisin allegedly enjoyed by the claimant or his ancestor or predecessor as having been enjoyed during the *time* of (*tempore*), here meaning the reign of, a particular, named king. Litigants or their legal representatives do this so regularly that we soon cease to notice that there is anything at all strange about it. But strange it clearly is, for it is not at all clear why this further specification should have been thought necessary and what relevance it could possibly have had: it is certainly not a particularly good way of being specific about when the seisin was enjoyed. The purpose, when

96 Cf. *Glanvill*, XIII, 19 (p. 161) with the earlier passage mentioning the assize (at IV, 1 [p. 44]).

97 *Pleas before the King or his Justices, 1198–1202, I*, ed. D. M. Stenton, Selden Society 67 (London, 1948), nos. 3497, 3533 and 3534. For a 1276 case where the defendant alleged that the last presentation to the the church alleged by the plaintiffs as their title had been in war-time (during the war between King John and his barons) rather than in peace-time during the reign of Henry III and said his ancestor had been the last to present in peace-time at the previous vacancy, see *Bartholomew of Calthorp and Alice of Repps v. Roger la Veylle of Bastwick*: PRO CP 40/15, m. 9.

98 PRO CP 40/91, m. 224.

it was introduced, which was probably during the reign of Henry II, for it is already a feature of the counts in the law-book *Glanvill*, may perhaps have been to ensure that claimants did not attempt to rely on the seisin of themselves or their ancestors or predecessors during the reign of King Stephen and thereby to make them rely instead on seisin in the reign of Henry I or Henry II, the only kinds of seisin that could establish or ground lawful title, in the eyes of Henry II and his advisers. But if this was the purpose the requirement had clearly outlived its original rationale by 1279 at the very latest when legislation enacted in 1275 finally achieved what had been intended, but not in fact achieved, by legislation of 1237, the barring of all claims that went back before the reign of Henry II. [99] By then it had clearly taken on a quite independent life of its own: a trap for those litigants who failed to mention it (as we know litigants did in 1204 and 1207)[100] and for serjeants who misstated the relationship of the king concerned to the present king (for this too was necessary in the case of former kings). John de Planaz was disavowed by his client in 1228 when he counted of his client's ancestor's seisin in the reign of a non-existent King Henry who was the grandfather of the king's father, King John.[101] By 1223 at the latest it had given rise to a still more curious development. It became possible (as it long remained possible) for a tenant in an action of right who denied the claimant's right to the property and the rightful seisin of the claimant or his ancestor,to pay a small sum (normally no more than half a mark) to secure a verdict of the grand assize (a jury of twelve knights) not just on the question of which of them had the greater right to the property at stake in the case but also, much more specifically, as to whether or not the claimant or his ancestor had been seised, as he claimed, in the reign of the specific king mentioned in the demandant's claim.[102] Later evidence indicates that the question about the time of the seisin was treated as a preliminary question. If the grand assize jury knew that the ancestor had not been or could not have been seised in the reign of the king concerned there was nothing more that they need say or do: that was enough to reject the claimant's claim, however good in all other respects his title might be.[103]

[99] P. Brand, '"Time out of Mind": The Knowledge and Use of the Eleventh- and Twelfth-Century Past in Thirteenth-Century Litigation', *Anglo-Norman Studies* 16 (1994), 37–54 (pp. 39–41).

[100] *Curia Regis Rolls*, III, 184; *Curia Regis Rolls*, V, 16.

[101] *Curia Regis Rolls*, XIII, no. 1194.

[102] *Curia Regis Rolls*, XI, no. 904. Something similar may be implied in the 1207 grand assize where the tenant put himself on the grand assize both *de jure et de tempore* (*Curia Regis Rolls*, V, 96–7) and in a 1212 grand assize where a proffer of two palfreys was made for grand assize on the question of greater right and whether the claimant's ancestor had been seised at the death of Henry I (*Curia Regis Rolls*, VI, 285–6).

[103] See Chief Justice Mettingham's instructions given to a grand assize in 1300 as reported in BL MS Add. 31826, fol. 323v; MS Add. 37657, fol. 48v; Oxford, Bodleian Library, MS Holkham Misc. 30, fol. 17v.

Limitation period

By the early thirteenth century the requirement that the claimant specify the reign in which ancestor or predecessor had been seised had also come to play another role: that of helping the courts to exclude claims that were based on facts that now lay too far in the past, claims that had lain unasserted for too long a period of time. Any count made in an action of right could not assert a seisin which had taken place prior to the 'limitation period' for this action. It looks as though the earliest limitation date in the action of right (applicable prior to *c.* 1200) allowed claimants to go back only as far as the reign of Henry I.[104] If, as seems possible, the introduction of this limitation date predates the availability of jury trial in the action of right, its original logic was perhaps nothing to do with the potential length of the memories of the jurors of the grand assize but was instead based on presumptions relating to trial by battle. The theory of trial by battle in the action of right was that the claimant's champion was fighting to support his attestation of facts he himself had seen or which his father had seen and had ordered his son to attest if ever a claim was made. Thus a cut-off period in the reign of Henry I may well have made sense in terms of the widest possible stretch of a two-generation memory.

The limitation date for the action of right was moved forward three times during the course of the thirteenth century, perhaps in accordance with the original logic but certainly with a view to continuing to exclude what had by then surely become stale claims. The first forward movement which seems to have taken place *c.* 1200 through legislation now lost barred claimants from going back further than the day of Henry I's death (the day that Henry was 'alive and dead'): and thus perhaps for the first time introduced claims in the action of right that were based (in theory at least) on the seisin of an ancestor or predecessor not just in the reign of a particular king but on a particular *day* in that king's reign.[105] Legislation of 1237 (which later came to be regarded as part of the Statute of Merton of 1236) attempted to move the limitation date for the action forward to the coronation of Henry II in 1154, thus allowing claims based on seisin in the reign of Henry II or afterwards, for any litigation initiated after Whitsun 1238,[106] but the courts adopted the view that it remained possible to bring litigation and assert a claim that was excluded by the legislation but allowed under the earlier legislation, provided the claimant was renewing litigation brought by an ancestor prior to 1237 if for some reason that litigation had not reached any conclusion. There therefore continued to be a trickle of cases in which claimants were allowed to base their claim on an ancestor's seisin in 1135

[104] Brand, '"Time out of Mind"', p. 41.
[105] Brand, '"Time out of Mind"', pp. 41–2.
[106] *Close Rolls, 1234–1237*, pp. 520–1.

down to 1279.[107] By then, the limitation date for the writ of right had been moved for the third (and last) time by *c.* 39 of the Statute of Westminster I (and to take effect only from two weeks after St John the Baptist 1276) to the beginning of the reign of King Richard (in 1189). This led to a flurry of cases based on seisins in the reign of Henry II being initiated that did not abate until 1291.[108]

Although there were a number of attempts during the course of the fourteenth century to secure a further movement forward of the limitation date in the action of right (and in other actions), none succeeded. This is, at first sight, a little puzzling, but may make slightly better sense if one remembers that by the end of the thirteenth century, the normal method of trial in the action of right had become the grand assize and that by then it had also become normal for the jurors of the grand assize no longer to be dependent solely on their own memories (which even in 1275 hardly went back to 1189) but instead to be presented with various kinds of written evidence such as charters, final concords and copies of enrolments from plea rolls. This written 'prompting' of the jurors' memories cut the necessary link with their own experience and meant that they no longer needed to be at least plausible witnesses of the facts on which they were called to decide. Once this had happened there was no longer any real need to move the limitation date forward.

The limitation date applied in the action of right did not feature in the writ itself. In this respect at least, it worked quite differently from the other two important limitation dates for legal actions: the limitation dates for the assizes of novel disseisin and mort d'ancestor. The question pre-formulated for the assize jury in the assize of novel disseisin was whether or not the claimant had been disseised of his free tenement by the defendant or defendants since a specific date (the limitation date). This limitation date was not a king's reign (as was normally the case with the writ of right), but rather a specific date in a king's reign. The earliest such limitation date of which we know is that given in *Glanvill* (written in the final years of the reign of Henry II) and it is the king's last voyage to Normandy.[109] Given the frequency with which Henry II crossed the Channel, this had the merit of allowing relatively regular updating without any need to change the wording of the writ. This strategy was not, however, followed by Henry II's successors who chose instead to use specific dates in their own reigns or the reigns of their immediate predecessors. Initially during the period down to 1237 the dates they chose were in the relatively recent past and were changed once each decade to keep them that way. This made it plausible that the jury really would know in general terms (if not to the nearest day) when

107 Brand, '"Time out of Mind"', p. 42.

108 Brand, '"Time out of Mind"', pp. 39–40.

109 *Glanvill*, XIII, 33 (p. 167).

the events concerned occurred.[110] In 1202 the limitation date for the assize became King John's coronation at Canterbury (at Easter 1201); in 1218 King John's last return from Ireland (24/26 August 1210); in 1229 Henry III's first coronation at Westminster (17 May 1220); in 1237 Henry III's first voyage to Brittany. It was at this point, however, that the practice shifted and it is not entirely clear why this should have been so. The limitation date for the assize of novel disseisin changed just once more (in 1275) as part of the changes introduced by c. 39 of the Statute of Westminster I when it was only moved forward (as from 1276) to the date of Henry III's first voyage into Gascony (8/9 May 1242).[111]

Something rather similar happened in the assize of mort d'ancestor. The question pre-formulated for the jury in the writ of mort d'ancestor was whether or not the father or other close ancestor of the claimant had died in possession of the land being claimed since the limitation date. The limitation date in this assize was from the beginning fixed in terms of a specific date and of a date in the rather more distant past than was the case with novel disseisin. The earliest limitation date of which we know (which was probably the limitation date from the time the action was created in 1176) was Henry II's first coronation (on 19 December 1154).[112] Around 1218/20 it was moved forward to the date of Richard I's first coronation (3 September 1189); and in 1237 to King John's last return from Ireland (24/26 August 1210).[113] Here too there was only one more movement forward, in 1275, under Westminster I, c. 39, and then only to the time of Henry III's coronation (17 May 1220).[114] Litigants certainly went on using both assizes through the fourteenth and into the fifteenth century, though there was then a marked decline in their usage, so the later failure to change the limitation dates does require some explanation. It looks as though the rather more generous limitation periods to which men had grown accustomed during the years prior to 1275 may have made it plain to them that there was really no great need for any explicit limitation dates in these actions at all for both assizes contained in their own factual requirements (the need for the persons disseised and disseising to be still alive; the need for the claimant to be a relatively close relative of the ancestor who had died in possession) what were in effect implicit limitation dates of their own.

In a process beginning perhaps as early as the later twelfth century, these three basic categories or forms of 'limitation date' came to be extended

110 For the evidence for these limitation dates, see the discussion by Hall in the introduction to *Early Registers of Writs*, ed. E. de Haas and G. D. G. Hall, Selden Society 87 (London, 1970), pp. xxxv–xxxvii.

111 *SR*, I, 36.

112 *Glanvill*, XIII, 3 (p. 150).

113 For the evidence for these limitation dates, see the discussion by Hall in *Early Registers of Writs*, pp. xxxv–xxxvii.

114 *SR*, I, 36.

outwards, to be applied to other forms of action or in other kinds of situation to which they were not originally applicable. The limitation date used in the assize of novel disseisin was adopted by statute in 1275 as the limitation date also for the action of *nuper obiit* (even though the latter was a claim by one co-heir against another for his share of their common inheritance based on the seisin of their common ancestor), probably to give a clear meaning to the requirement that the ancestor should have died recently (*nuper obiit*) but this was probably not the earlier rule: it did not appear in the writ itself.[115] The limitation date used in mort d'ancestor was also by 1229 at latest being used in the action of naifty (where it did appear in the writ): the lord having to assert that he had been in possession of the villein whom he was claiming within the limitation period. By 1237 the same limitation period was also apparently applicable in most (but not all) writs of entry (claiming land on the basis of a single flaw in the tenant's title), where the demandant had to show that he or the ancestor from whom he derived his title to the land had been in possession of the land being claimed within the limitation period, though this requirement was not incorporated into the formulas of any of the various forms of writ of entry.[116] During the same decade it came to be applied to the new actions of aiel and cosinage which allowed (like mort d'ancestor) claims to be made based on the seisin of an ancestor at the time of his death but extended the principle to more distant kinsmen (and again here the writ formulas that were developed said nothing of this requirement).[117]

A more ambitious leap outwards was made by statute (Westminster II, c. 2) in 1285. Hitherto, the general rule followed in regard to the use of distraint to enforce the performance of services by a tenant to his lord was that this could only be justified, when challenged in the action of replevin, if the lord could show that he had himself been in receipt of the services or an immediate ancestor had died in receipt of them. The legislation substituted a much more generous rule: one that allowed a lord to justify his distraint if he could show that he or any of his ancestors had been in possession of them at any time since the limitation date in novel disseisin.[118] Here the limitation date is no longer being applied to a form of action being used to initiate litigation, but for the first time as a limitation on the use of an approved form of self-help. The Statute of Quo Warranto of 1290 (as applied in practice) brought a different kind of extension by analogy of the use of the limitation date in the action of right when the king accepted that possession of franchises since the reign of King Richard (the limitation date in the action of right) would in

[115] *SR*, I, 36; *Early Registers of Writs*, CC 186. The earlier rule seem to have required the ancestor to have died within some fixed period of time (seven or fourteen or twenty years): *Brevia Placitata*, pp. 9–10.

[116] *Close Rolls, 1234–7*, pp. 520–1.

[117] *Bracton's Note Book*, ed. F. W. Maitland, 3 vols. (Cambridge, 1887), III, 228 (pl. 1215).

[118] Brand, *Making of the Common Law*, pp. 292–5.

future be accepted as a legitimate title to the exercise of franchises, as a legitimate defence to any challenge by the king's action of *quo warranto* though only if that exercise had been continuous.[119]

This seems in turn to be related to a more widespread phenomenon (or set of phenomena) that can be witnessed taking place during the course of the thirteenth century: a growing acceptance that certain types of right or entitlement could only be grounded in continuous (and peaceable) possession since 'time out of mind' or might be grounded in such possession in the absence of a written title, and a growing identification of the limitation period in the action of right as marking the point beyond which 'legal memory' did not go: so that continuous possession since time out of mind came to be identified with continuous possession since the limitation period in the writ of right. A good example of the kind of entitlement that could only be grounded in continuous possession since time out of mind was the right claimed by the lords of the Cumberland barony of Gilsland that no free tenant within the barony could 'approve' waste land without their consent (asserted in a case of 1291)[120] or the right claimed by the master of the Templars to a share of the chattels of all their dead free tenants (asserted in a case of 1294).[121] Good examples of entitlements that could be grounded in a written grant but might alternatively be grounded in continuous seisin since time out of mind were a tenant's right to be acquitted by his lord of additional services claimed by their overlord and the right to warranty from a lord on the basis of the performance of homage alone. 'Time out of mind' seems to have been a concept borrowed from the learned law but was already being identified with the limitation period in a case of 1247 where Warner fitzEudo pleaded as his title to common in the lands of the abbot of Kirkstead that he and all his ancestors had been seised of common 'since time of which there is no memory and from such a time that no writ of right runs' ('a tempore de quo non constat memoria et a tali tempore quod non currit aliquod breve de recto'), namely during the reign of Henry I and in the reigns of Henry II, Richard, John and the present king.[122]

But the limitation date for the writ of right was not as yet by 1300 literally the limit of 'legal memory', the point beyond which it was impossible for litigants or the courts to go in the course of pleading, or was only just as yet becoming that limit.[123] Litigants (particularly old-established religious houses) regularly produced pre-1189 (indeed, pre-1066) charters in court and it long remained the practice to prove ancient demesne status by

[119] D. W. Sutherland, *Quo Warranto Proceedings in the Reign of Edward I* (Oxford, 1963), pp. 91–7.

[120] *Walter of Rothbury and his wife Isabel v. Maud of Moulton*: PRO CP 40/91, m. 302.

[121] PRO CP 40/104, m. 105.

[122] PRO JUST 1/81, m. 6.

[123] Brand, '"Time out of Mind"', pp. 47–51.

reference to Domesday Book. Reference to facts that had occurred prior to 1189 was often necessary in wardship cases where two claimants to the wardship of an heir's person were attempting to establish which of their ancestors had been first to enfeoff the heir's ancestor. References to a relatively distant past were also a common feature of actions of *monstraverunt* brought by the villein sokemen of ancient demesne manors against their lords, claiming to revert to the fixed services they had enjoyed while their manor was in the hands of one of the king's ancestors, sometimes as distant as William the Conqueror.

CONCLUSION

The justices, clerks and lawyers of late medieval England developed their own distinctive ways of dividing up, and describing, the passing of time in the English royal courts, a system which seems to have served their needs even if its precise rationale is not always wholly transparent; and a system whose workings need to be understood by any historian wishing to understand the records which that legal system produced. The same group of legal specialists also created their own distinctive ways of handling what was already for them the past: different rules and practice about how detailed a specification needed to be given in litigation of when particular past events had occurred (and the appropriate ways of expressing this); developing distinct categories of past time based on some of their more important characteristics (times of peace and times of war) and applying different rules to them; developing rules about how far into the past litigants might go when making different kinds of claim and about the creation and recognition of rights by the lapse of time when particular customary rights had been regularly exercised or gone unchallenged. The starting point of lawyer's time was always the time of the real world and lawyer's time never entirely lost contact with real time, but the way in which it developed reflects the growing sophistication of that system and its creation of its own distinctive ways of thinking and arguing about the past, and doing and recording business in the present.

5

Time and Urban Culture in Late Medieval England

Chris Humphrey

The widespread introduction of mechanical clocks into Europe in the fourteenth century laid the foundations for our modern sense of time. But how rapidly was this process accomplished, and what happened to the older systems of time-keeping that were supplanted? This study draws on evidence from medieval York in order to examine the nuances and impact of this shift in the nature of daily time in the Middle Ages. York's case suggests that the emergence of the 'equal-hours' system as the dominant form of time-keeping in the city owes less to the benefits of new technology, and more to the aspiration by a group of citizens to organize their social life in terms of a distinctive daily programme. This observation highlights the need to think beyond the simple adoption of mechanical clocks as the main factor in the development of a modern equal-hours day: after all, equal-length hours were around a long time before clocks appeared on the scene.[1] The essay begins with some preliminary remarks about approaches to the history of time, and discusses two principal senses of daily time in the Middle Ages. It then moves on to consider the experience of daily time in the late medieval English town, with particular reference to evidence from the city of York. The conclusion puts the case for a more sophisticated understanding of the relationship between towns and the birth of clock-time, including a consideration of the role of cultural factors in the creation of novel forms of temporality.

NARRATIVES OF TIME

One approach to the history of time-keeping has been to collect references in the historical record to different time systems, plot their changes, and thereby show shifts such as the change from the Julian to the Gregorian calendar,

[1] 'It should not be thought that the concept of equal hours ("horae de clock", in the late fourteenth century), had to wait for the invention of the mechanical clock, as is sometimes suggested.' J. D. North, 'Monasticism and the First Mechanical Clocks', in *The Study of Time II*, ed. J. T. Fraser and N. Lawrence (Berlin, 1975), pp. 381–98.

monastic hours to clock hours, or the AD dating system replacing mixed year-dates. While providing an account of how perceptions of time have changed historically, the limitation of this kind of overview is that it fails to capture the way that people related to time as individuals or as members of networks and institutions in past societies. To obtain such a perspective, time-schemes need to be thought of as more than just abstract systems of measurement: they need to be seen as narratives or ways of ordering experience that are of shared significance to a group or culture. This idea can give us a fresh series of insights into the numerous attempts that were made to reorganize, or even 're-narrativize', the structure of the day in England in the late Middle Ages.

It is well known that there were two main public means of dividing up daily time in medieval England. In many parts of the ancient world and up until the fourteenth century in Europe, the day was organized according to the temporal hours system. Under this convention each hour is variable in length, because it is calculated by dividing the period of daylight into twelve parts, and the same at night. A daytime hour in the summer was therefore longer than its winter counterpart. The numbering of these hours began at first light, giving prime, terce, sext and nones at the main hours of the day. The alternative and newer system was to use equal hours, creating twenty-four units of time that were all of the same length. Although equal hours were clearly known about in the early Middle Ages, as shown for instance by wax candles marked with equally-spaced intervals, they did not have a central role in public time-keeping.

It is also well known that equal hours came to replace temporal hours as the dominant medium of daily time in the Middle Ages. That said, the situation on the ground was a good deal more complex than terms like 'Church's time' or 'merchant's time', as popularized by Le Goff, may suggest.[2] In fact what is striking about England in the fourteenth and fifteenth centuries is the sheer variety of ways of organizing life at the level of the everyday that can be discerned. The ordinances regulating the work of the masons and other workers at York Minster in the later fourteenth century illustrate this diversity well:

> It is ordained by the chapter of the church of St Peter of York that all the masons that shall work on the works of the same church of St Peter shall from Michaelmas Day unto the first Sunday of Lent be each day in the morning at their work, in the lodge that is ordained to the masons to work in with the close beside the aforesaid church, as early as they may see skilfully by daylight for to work. And they shall stand there truly working at their work all the day after, as long at they may see skilfully to work, if it

[2] Le Goff, *Time, Work, and Culture*, pp. 29–42. Le Goff admits that his essay 'has no other purpose than to stimulate a more intensive study of a history which raises numerous problems' (p. 41).

> be all workday; otherwise till it be high noon smitten by the clock when holy day falls at noon, so that it be within the aforesaid time between Michaelmas and Lent. And in all other time of the year they may dine before noon, if they will, and also eat at noon when they wish, so that they shall not stray from their works in the aforesaid lodge at any time of the year at dinner time, but such a short time that no skilful man shall find fault in their absence. And at time of meat, at noon, they shall, at no time of the year, stray from the lodge nor from the work aforesaid, more than the space of the time of an hour, and after noon they may drink in the lodge. And for their drinking time between Michaelmas and Lent they shall not cease nor leave their work passing the time of half a mile away.[3]

This passage makes use of several different schemes for reckoning daily time, including sunrise, estimates of the amount of daylight in which it is reasonable to work, estimates of how long it is appropriate to be away from work, equal hours as measured by a clock, and measures of activity such as the amount of time that it takes to walk a particular distance. Clearly, rather than indicating an absolute divide between temporal hours and equal hours, a document such as this illustrates how a range of time indications was still in practical use after mechanical clocks had made their first appearance. Medieval patterns of daily life appear for the most part to have been organized according to tradition, expediency, practical concerns and compromise, as opposed to a desire to adhere slavishly to any one system of time reckoning.

On the other hand, it is also important to recognize that across a range of different sources, a day or period of twenty-four hours that was ordered by a single temporal system did offer a particular kind of aspirational ideal in the medieval period. A good example is the three-fold division of the day into periods of eight hours, attributed to King Alfred by William of Malmesbury. A popular idea in the Middle Ages was that King Alfred had divided his day into three equal parts, one for writing, reading and praying, the second for caring for his body and the third for dispatching the affairs of the realm. Later chroniclers had Alfred burning candles in order to measure these periods of time.[4] This story provides an example of what can be called an 'aspirational narrative of the day', that is, an idealized scheme for ordering daily life that is built around only one kind of time-signal, in this case the measured candle. There are many instances of these idealized, repeatable and public schemes for structuring time that can be identified in medieval England. The liturgy is the most obvious example, closely followed by the monastic rules that allocated time to particular activities during the day and night. Less well known but none the less important were literary examples such as the Hours

[3] 'Rules for the conduct of the Masons of York Minster', in *English Historical Documents 1327–1485*, ed. A. R. Myers (London, 1969), pp. 1182–4 (p. 1183) (adapted).

[4] G. Langenfelt, *The Historic Origin of the Eight Hours Day: Studies in English Traditionalism* (Stockholm, 1954), pp. 26–37.

of the Passion lyrics, which recalled the times of the day at which Christ suffered torment and death on Good Friday.[5] Fragments of idealized temporal schemes also occur in works such as Geoffrey Chaucer's *Canterbury Tales*: in the 'Shipman's Tale', for instance, the monk who is awake early and walking in the garden observes that 'it oghte ynough suffise/Five houres for to slepe upon a nyght', except for old and married men (ll. 1290–1).[6]

Idealized views of time also occur in medieval parliamentary statutes, as part of an effort to structure the day for groups such artisans and labourers. These groups were perceived to have problems when it came to organizing their working time in a way that was amenable to their employers. Thus in 1495 and again two decades later, attempts were made to regulate the hours that artisans and labourers worked. The complaint was that these groups wasted much of their day and did not deserve their wages, because they were late to work, went home early, took a long time at breakfast and dinner, and slept too much in the afternoon. They were therefore instructed to be at work between five o'clock in the morning and remain there until between seven and eight o'clock in the summer, and to work between sunrise and sunset in the winter (admittedly, temporal and equal hours are mixed here).[7] All of the schemes considered so far set out a means of structuring one particular day that is also implicitly *every* day: that is, all subsequent days can and should be ordered in this manner too. So we have two different ways of thinking about time represented here: what we might call 'practical time' and 'aspirational time'. On the ground, time was organized according to a range of temporal indicators, as the schedule drawn up to regulate the work of the minster masons makes clear. On the other hand, literary and legislative sources reveal some idealized views of time, based around a unified way of measuring it such as the candle or the clock.

The differences we have observed prompt us to consider the following interesting question. In what circumstances, and for what reasons, did particular aspirational conceptions of daily time prove useful, desirable or attainable in the Middle Ages? This question is especially relevant for institutions whose task it was to provide public, future-oriented, prescriptive narratives for governing the lives of individuals and groups. Institutions having this responsibility to a greater or lesser degree included the Church and monastic houses, the state, urban communities and employers of wage labour. To ask the same question in a different way: when and why did institutions or groups attempt to reform or even create from scratch exemplary 'narratives of the day' that were organized by reference to only one time-signal? Answering such a question for the medieval town can help us to

[5] See, e.g., 'The Dolours of Our Lady', in *Religious Lyrics of the XVth Century*, ed. C. Brown (London, 1939), pp. 138–40.

[6] *Riverside Chaucer*, p. 204.

[7] *SR*, II, 586–7; III, 125–6.

understand what was at stake in the acquisition and use of clocks by urban communities, or at the least the authorities acting on their behalf. In turn the shift from the temporal-hours system to the equal-hours system can be explained in a way that does not assume the innate superiority of clock-time: cultural and political factors also need to be taken into account. A case study of the city of York in the fourteenth and fifteenth centuries will enable us to explore this issue at ground level for a particular community.

DAILY TIME IN THE MEDIEVAL ENGLISH TOWN

When we consider the subject of time in towns in fourteenth- and fifteenth-century England, the evidence throws up an apparent paradox. On the one hand, clocks have been seen by those working on the history of time-keeping as symbols of urbanity, bound up with the rise of urban communities as autonomous, self-regulating bodies.[8] Although this view is based for the most part on continental sources, there having been much less work on the English medieval evidence to date, a similar case can be made for this association in England too. We can think, for example, of the fifteenth-century text *Dives and Pauper*, where it is observed that 'in cities and towns men rule themselves by the clock'.[9] On the other hand (and this is where the paradox lies), town governments and groups of citizens in medieval England were slow to get hold of and use clocks in comparison with the other institutions in their midst such as cathedrals, university colleges and monasteries. In fact, towns were not exactly at the cutting edge when it came to the acquisition of mechanical clocks, since chronologically at least they lagged somewhat belatedly behind the Church and the crown. Given this situation in the evidence, we might legitimately ask how urban life came to symbolize a clock-based way of living at all, both in the period itself and subsequently to historians of time and towns. The answer must lie in the relationship between clocks and a particular narrative of the day in towns, rather than just the simple use of clock hours themselves. At stake are the practical uses to which clocks were put, and the kinds of identity that a new temporal scheme for dividing up the day could create.

In order to investigate these issues, the second part of this study will examine evidence relating to time and time-keepers in the medieval city of York. York's three main public clocks and the public role that these time-keepers had will be considered in a survey of the evidence that provisionally collates information from a number of sources. In the conclusion, the ideas about time and narrative raised at the beginning of the essay will be

[8] See G. Dohrn-van Rossum, *History of the Hour: Clocks and Modern Temporal Orders*, trans. T. Dunlap (London, 1996), pp. 4–15, and particularly the quotations given on p. 13.

[9] *Dives et Pauper*, ed. P. H. Barnum, 2 vols., EETS 275 and 280 (Oxford, 1976–80), I, 120.

used to offer some new insights into the urban origins of a clock-based division of the day.

St Mary's Abbey

St Mary's Abbey would have followed the Rule of St Benedict in ordering its day, celebrating a series of religious offices at the times provided by the temporal-hours system. Given the wealth and power of the abbey, it is not surprising to find it acquiring a clock early in the fourteenth century. The abbey's chronicle records that in 1324 Stephan de Austerwick, then sacristan, had a new *orologium* and bell made.[10] *[H]orologium* is a generic Latin word that can denote a number of different time-keeping devices such as sundials, water clocks or mechanical clocks.[11] In this case, it is likely that it refers to a mechanical clock, since a number of other cathedrals and monastic institutions in England were acquiring such clocks at this time. For instance, the famous Norwich clock was built by Roger Stoke between 1321 and 1325, and at Lincoln Cathedral in 1324 the treasurer made a donation for a new clock because the cathedral did not possess what most other cathedrals, churches and convents had.[12]

What is interesting about St Mary's Abbey is that we can gauge some of the effects of the installation of the clock. By the late fourteenth century the times at which the abbey's daily round of services were celebrated were determined in part by clock-time rather than in relation to daylight hours alone. For instance, the abbey's custumary ordered that the office of prime, the first service of the day and originally celebrated at dawn, should begin when the clock struck seven at all times of the year.[13] The significance of this shift was that the abbey's services were no longer synchronized with a twelve-fold division of night and day: instead they were to begin at an abstract signal determined by the clock. The severing of the link between dawn and the first service of the day represents a crucial first step in the demise of a traditional system of time-keeping that was based around the observation of religious services at the corresponding hours of daylight. Nonetheless, this example also shows that two different systems were in operation simultaneously, with the clock-hour system providing the cue for a temporal-hour cycle that presumably occupied the rest of the day.

The abbey's clock and bells do not appear to have been used as a wider

10 *The Chronicle of St Mary's, York*, ed. H. H. E. Craster and M. E. Thornton, Surtees Society 148 (Durham, 1934), p. 73.

11 Dohrn-van Rossum, *History of the Hour*, pp. 52–3.

12 See the discussion of medieval cathedral clocks by F. C. Haber, 'The Cathedral Clock and the Cosmological Clock Metaphor', in *The Study of Time II*, ed. Fraser and Lawrence, pp. 399–416.

13 *The Ordinal and Customary of the Abbey of Saint Mary, York*, ed. L. MacLachlan and J. B. L. Tolhurst, Henry Bradshaw Society 73 (London, 1936), p. 49.

reference point in medieval York, except in one case. In an earlier set of rules governing the working times of the masons and other workers on York Minster, *c.* 1355, their finishing time at the end of the day was to be indicated by the bell of St Mary's.[14] It is not clear why one of the abbey's bells was used rather than one belonging to the cathedral itself. This is the only known example of the abbey's involvement in a sense of public time in medieval York, although presumably many people within the abbey's enclave used its bells and other time-indicators as temporal reference points.

York Minster

As with St Mary's, the daily round at York Minster was organized according to the religious offices that were to be celebrated that day. The minster's bells provided a reference point for internal services and also seem to have featured as a point of reference for the wider public life of the town. For instance, in a set of ordinances regulating the Fishmongers' Guild from 1301, it was ordered that no fish was to be sold after vespers was struck at the church of St Michael on Ouse Bridge, until prime was struck at York Minster the next day.[15] This example illustrates the temporal-hours system in action, and indicates a common feature of much early time indication in the way that it refers to the actual time-keeper as well as to the time of day.

Later in the fourteenth century, the minster invested in some new time-keeping equipment, including a new clock from John Clareburgh, and a new 'great bell' to replace the one broken in 1360.[16] The size and location of this new clock is unclear, although antiquarians such as Torre and later Drake make reference to a 'clock house' which stood to the left of the inside of the south transept door.[17] This position matches the location where the early sixteenth century striking jacks were located on the outside of the south transept.[18] A small belfry on the roof of the south transept housed the clock bell, and this belfry is illustrated in early pictures of the minster (see Plate 1). Given that the body and tower of the minster underwent successive phases of rebuilding in the late Middle Ages, it is difficult to see where the main ring of bells could have hung for any length of time, prior to their installation in the south-west tower in the mid-fifteenth century.[19] This observation lends support to Christopher Norton's intriguing thesis that the name of the church adjacent to the minster, St Michael-le-Belfry, in fact reflects its

[14] *The Fabric Rolls of York Minster*, ed. J. Raine, Surtees Society 35 (Durham, 1859), pp. 171–3.
[15] *York Civic Ordinances, 1301*, ed. M. Prestwich, Borthwick Papers 49 (York, 1976), p. 13.
[16] J. Browne, *The History of the Metropolitan Church of St. Peter, York*, 2 vols. (London, 1847), I, 168–70; D. E. Potter, *The Bells and Bellringers of York Minster* (York, 1987), p. 8.
[17] York, York Minster Library, Torre Manuscript LI/7/1, fols. 30–30v.; F. Drake, *Eboracum: or the History and Antiquities of the City of York* (London, 1736; repr. East Ardsley, 1978), p. 529.
[18] Potter, *Bells of York Minster*, p. 58.
[19] Potter, *Bells of York Minster*, pp. 8–10.

Plate 1 Francis Place, etching of the west front of York Minster (1730). Reproduced by permission of York City Art Gallery

status as the minster's original belfry.[20] The acquisition of a new clock by the minster changed the way that daily, public time could be measured in the city. As we have already seen, as early as the 1370s the masons were finishing work on the eve of a feast day when noon was rung by the minster's clock. By the early fifteenth century we can find the minster's clock being used as a wider point of reference in the city. For instance, the ordinances of the Girdlers' Guild, dated 1417, ordered that no-one of the craft should work on a Saturday after twelve o'clock was struck at York Minster, on pain of a fine of a pound of wax.[21] This enabled the afternoon to be observed as the eve of Sunday.

These examples help to illustrate how the minster's bells were used as temporal referents in the medieval city: clearly when its clock was acquired this added another layer of time for organizing rhythms of work or other activities. Our main impression again is of a mixed system, in which the more subjective senses of time based upon daylight hours were used alongside clock-time.

The city of York

The government of the city of York sought to provide public time-keeping facilities in areas where all citizens had an interest. The fact that these efforts were taking place when the city was in the process of acquiring a certain degree of autonomy indicates the close relationship between civic identity and a sense of corporate time at this period.

An early reference to the city's own time-keeping measures helps to illustrate this relationship between time and civic identity. In 1380 a parliamentary petition complained that the legitimate mayor of the city, John de Gysburn, had been expelled and another man, Simon de Quixlay, had been forced to take his place. This was against the law and good customs of the city, and against the wishes of its 'good men', who had been compelled to swear loyalty to the new mayor by the usurpers. Not only that, but the usurpers are alleged to have made a new ordinance which declared that whenever the bells on the bridge sounded *aukeward*, day or night, all the commons of the city should rise up and proclaim new ordinances. There are several interesting points about these claims. The presumption must be that the bridge is Ouse Bridge, suggesting that by this date the city had installed a set of bells for indicating particular times of the day there. The insertion of the Middle English word *aukeward* in a passage otherwise written in French is also of interest, helping to convey a sense of dissonance in relation to what

20 C. Norton, 'The Anglo-Saxon Cathedral at York and the Topography of the Anglian City', *Journal of the British Archaeological Association* 151 (1998), 1–42 (p. 5).

21 *York Memorandum Book*, ed. M. Sellers, 2 vols., Surtees Society 120 and 125 (1912–15), I, 184.

was formerly a harmonious situation. Overall, the petition gives the sense that there was already an established way of ringing these bells, and that altering this regime was a challenge to an ordered sense of public time, the 'good custom' that is referred to.[22]

The common point of reference provided by the bells on Ouse Bridge, adjacent to the council chambers there, was further enhanced by the city's acquisition of a clock some time in the late fourteenth or early fifteenth century. This was also installed on Ouse Bridge, in the twelfth-century chapel of St William (see Plate 2). The civic records do contain some late fourteenth century references to particular hours of the day, suggesting that the clock-hour system was already in use by then. That said, the first definite date for the use of the Ouse Bridge clock as a temporal reference point was in 1428. In that year there was a complaint in the city that the butchers were not opening their shops for business on Sundays. Therefore it was ordered that butchers should keep their shops open on Sundays, until eight of the bell of the clock of the commonalty on Ouse Bridge. They were then to keep their shops closed until after vespers at their parish church, when they were to open them again.[23] This is an interesting example both of a city-wide sense of time which is the responsibility of the city, and a sense of local time that cuts across it. Both clock-time and temporal hours were in use simultaneously.

In the course of the fifteenth century there are several mentions in the city's accounts for payments towards the clock's upkeep, including payments to John Ripley, who turns up at Ripon Minster a few years later, mending the clock there.[24] Once into the fifteenth century, references to clock-time in the civic records become much more numerous (see below).

CONCLUSION: CLOCKS AND THE NEW TEMPORAL NARRATIVE

Given the evidence that we have considered so far, we may conclude that York's civic government was fairly slow in getting hold of its own clock, when compared with the other institutions in the city, and if we are using the criterion of the earliest historical reference. This conclusion does not fit with the view that we have from other sources and accounts, which is that town communities were closely associated with temporal innovation, or even that towns were the pioneers of clock-time. We may therefore ask whether York was just an exceptional case, or if there is something more to be said here? In fact, some further reflection can help us to get a better sense of how town bodies and clocks became actually and symbolically linked in

22 *RP*, III, 96.

23 *York Memorandum Book*, ed. Sellers, II, 183.

24 *York City Chamberlains' Account Rolls 1396–1500*, ed. R. B. Dobson, Surtees Society 192 (Durham, 1980), pp. 26 and 35.

Plate 2 I. Halfpenny, etching of the Ouse Bridge and St William's Chapel, York (1807). Reproduced by permission of York City Art Gallery.

this period. If we consider York Minster and St Mary's Abbey, their acquisition of clocks was a further means of shaping the day in those institutions, indicating when the service of prime began for instance. Crucially the main structure of the day remained intact: the liturgy and its daily services were still the pre-eminent means of dividing up the day, based around liturgical texts and services. In fact, liturgical texts were literally 'books of hours', since reciting prayers and carrying out services gave each day its structure and rhythm.

By contrast, when York's civic government acquired its clock, the impact on the pattern of urban daily life was much greater, opening up new possibilities for a developing sense of corporate time. Back in the earlier fourteenth century, trade activities that had a distinctively 'common' aspect to them, such as the times at which fish might be bought and sold, appear to have been regulated by reference to local time-keepers such as the bells of York Minster. An important subsequent development was the establishment of common bells on Ouse Bridge and Foss Bridge, since this gave the citizens the ability to ring times of common interest themselves. This shift marks the beginning of what we might call an 'institutional day', in which a collectivity provides a structured pattern of time signals for its members to refer to. What is interesting about the city's subsequent acquisition of a clock is that, instead of using it as an adjunct to the temporal system already in use (as the ecclesiastical institutions had done), the civic clock provided the basis for an entirely new narrative for the day, that of equal hours. Temporal hours were not gently retired but actually imagined out of existence by the proponents of a new corporate time based upon equal hours.

This is where the real heart of the link between towns and clocks lies: it is not just about which institutions first acquired mechanical clocks and when by implication they began to use them. Rather, the critical issue is the way that groups of citizens like those in York aspired to rewrite the day around a different narrative altogether, one that was 'of the clock' and signalled by their *own* clock. Here was a means of creating a new 'mean time' that was public and city-owned, both for the practical purpose of organizing daily life and as symbolic of a distinctive urban identity. The novelty of this effort is perhaps hard for us moderns to grasp, since the clock-time system seems a natural and transparent way of organizing our everyday experience. However, we might suggest that, at least initially, clocks were never envisaged as the basis for a complete and public system of ordering daily life. Indeed, as we have seen, for the most part the evidence points to their use in a more circumspect way, as adjudicators of particular points in time, for making astronomical calculations or just as spectacular demonstrations of wealth and power. To envisage the sequence of clock hours as the basis for a whole new system of public organization, and to put this scheme in practice, is quite another matter entirely. As a feat of imagination it arguably represents as much of a breakthrough as the invention of the mechanical escapement for clocks, an event that too many commentators have extolled as the principal temporal innovation of the age.

Telling the time in an entirely new way is not an idea that the citizens of medieval York can take all the credit for, since it is clear that town bodies right across Europe were adopting clock hours at this period.[25] What was at stake in England, though, is the issue of urban autonomy, and this particular development in time-keeping seems to fit well with what else we know about the struggles of English towns to achieve self-governance and self-definition in this period. In fact, in York's case the link between clock-time and civic identity is shown quite clearly in the later fifteenth century, when the records are more numerous and more specific about the daily timing of events. In these later records, references to temporal hours are rare: instead, events that have a public and communal aspect to them are given on the whole in clock-time and, more specifically, in relation the city's own clock on Ouse Bridge. The city's clock was certainly used as a point of reference in a wide variety of activities, from the times at which the gates and posterns of the city should be sparred, and the times of the watch, to the times when bread should be sold at Thursday market.[26] Evidently this mode of comprehending the day had become the principal medium in which business of a civic or common character was conducted and articulated.

To sum up, our modern understanding of 'o'clock' as a simple contraction overlooks the extensive phrasing used at an earlier period, when times were given 'of the bell of the clock of the commonalty on Ouse Bridge'. This was time as heard from a particular place, and in its earliest use it was partisan and a conscious act of self-definition. This paper has shown how the shift from a temporal-hours system to an equal-hours one was brought about by more than just the inevitable march of a new technology. In fact, the equal-hours day owes its existence to the determined individuals who imagined a distinctive and original narrative for their activities that was 'of the clock'. It is this powerful temporal narrative that continues to shape our conception of the nature of daily time in the modern world.

[25] See Dohrn-van Rossum, *History of the Hour*, chap. 5, 'From Prestige Object to Urban Accessory: The Diffusion of Public Clocks in the Fourteenth and Fifteenth Centuries'.

[26] *The York House Books 1461–1490*, ed. L. Attreed, 2 vols. (Stroud, 1991), II, 690; *York Memorandum Book*, ed. Sellers, II, 170–2.

6

In Search of Lost Time: Missing Days in *Sir Cleges* and *Sir Gawain and the Green Knight*

Ad Putter

In his book *Categories of Medieval Culture*, A. J. Gurevich writes: 'Man is not born with a "sense of time"; his temporal and spatial concepts are invariably conferred upon him by the culture to which he belongs.'[1] That the medieval sense of time differed from our own is something that medievalists discover the hard way, when, for example, they struggle to date a charter because the medieval scribe omitted the date or, more often, because he recorded it in a notation that bears an obscure relation to our modern one. Books have been written to help us convert medieval currencies of time into our own and avoid the mistakes that occur when incompatible currencies are interchanged.[2] R. L. Poole tells the cautionary tale of a historian who, knowing Thomas Becket to have died on 29 December 1170, believed the medieval chroniclers to be in error when they dated it to 1171. Unfortunately, the error was his, because in that period the new year began on Christmas Day and not on 1 January, as became customary after *c.* 1250.[3]

Since time also passes in the fictional worlds of medieval literature, literary critics risk making comparable mistakes when they impose their own sense of time on that implied by the medieval text. I say 'implied' because in literature the only time that really passes is the time spent reading or listening (the 'time of narration'): 'narrated time' is an illusion and passes only by suggestion.[4] We arrive at an imaginary chronology by supplementing the temporal markers in the text with our own experience of the world. And the possibility of error arises when this experience that we draw upon to restore the fullness of time conflicts with the experience originally envisaged by the author.

This essay discusses time in two later Middle English romances, *Sir Cleges*

[1] A. Gurevich, *Categories of Medieval Culture*, trans. G. L. Campbell (London, 1985), p. 27.

[2] For a recent example and further references see *Handbook of Dates for Students of British History*, ed. C. R. Cheney, rev. M. Jones (Cambridge, 2000).

[3] A. L. Poole, 'The Beginning of the Year in the Middle Ages', in A. L. Poole, *Studies in Chronology and History* (Oxford, 1934), pp. 1–27 (p. 1).

[4] The terms 'narrated time' and 'time of narration', translating German *erzählte Zeit* and *Erzählzeit*, are taken from M. Bal, *Narratology: Introduction to the Theory of Narrative*, trans. C. van Boheemen (Toronto, 1985).

and *Sir Gawain and the Green Knight*.[5] Both romances are set at Christmas and both present an interesting problem of chronology: while they supply a number of temporal markers, the calendar that can be reconstructed on the basis of these markers does not work out. A day appears to be missing. I hope to show that in *Cleges* the chronological error is not the poet's but that of his editors, who have failed to adjust themselves to the medieval rhythms of Christmas that the poet took for granted. In the case of *Gawain*, too, a neglected dimension of the medieval experience of Christmas has a bearing on the problem of the missing day, though here the historical context, rather than resolving the problem, only makes things worse. My evidence supports the hypothesis that the error is scribal. After showing why the standard textual emendation cannot be correct, I shall propose a different solution.

SIR CLEGES

First to *Cleges*. This tail-rhyme romance, probably composed *c.* 1370–80,[6] tells the story of a knight, Sir Cleges, who has fallen on hard times and is obliged to mortgage one estate after the other to maintain his high standards of hospitality at Christmas. As another Christmas approaches and Cleges is down to his last manor house near Cardiff, where he lives with his wife Clarys and his children, he falls into despair, but Clarys cheers him up and makes him accept God's will with good grace. After mass, Cleges goes alone into his garden and prays to God, thanking him for all that has been ordained for him. He then notices that the cherry tree is in leaf and laden with fruit. Clarys assures him that this is a good omen and suggests that Cleges present a basket of cherries to the king, who holds court in Cardiff Castle. Accompanied by his son, Cleges goes to find the king, but he is stopped by three household officers (the porter, the steward and the usher), who will not let him pass unless he promises to give each a third of his eventual reward for

[5] I have used the most recent edition of *Sir Cleges* in *The Middle English Breton Lays*, ed. A. Laskaya and E. Salisbury, Consortium for the Teaching of the Middle Ages (Kalamazoo, MI, 1995). References to *Gawain* are to the edition by J. R. R. Tolkien and E. V. Gordon, rev. N. Davis (Oxford, 1967).

[6] The romance survives in two fifteenth-century manuscripts: Edinburgh, National Library of Scotland, Advocates MS 19.1.11, and Oxford, Bodleian Library, MS Ashmole 61. Both manuscripts are described in G. Guddat-Figge, *Catalogue of Manuscripts of Middle English Romances* (Munich, 1976). On Ashmole 61 see further the articles by L. S. Blanchfield: 'The Romances of Ashmole 61: An Idiosyncratic Scribe', in *Romance in Medieval England*, ed. M. Mills, J. Fellows and C. M. Meale (Cambridge, 1991), pp. 65–87; and 'Rate Revisited: The Compilation of Narrative Works in MS Ashmole 61', in *Romance Reading on the Book: Essays in Medieval Narrative Presented to Maldwyn Mills*, ed. J. Fellows, R. Field, G. Rogers and J. Weiss (Cardiff, 1996), pp. 208–20. For reasons I cannot go into here, I would place *Sir Cleges* not, as often argued, *c.* 1400, but a couple of decades earlier.

the cherries. When Cleges finally reaches the king with his offering, he is offered a reward of his own choosing. Cleges asks for twelve blows, which he proceeds to divide amongst the three greedy officers. In the meanwhile, a minstrel sings to the king the lay of Cleges. The king remembers his once-favoured knight and wonders what has become of him. Cleges returns to the king to thank him for his gift and explains to everyone's amusement why he asked for twelve strokes in return. When he reveals his name he is lavishly remunerated and appointed steward. Cleges and Clarys end their days on earth comfortably and go to heaven thereafter.

As criticism of the poem has shown, the various motifs of the story (the Spendthrift Knight, the Unseasonable Growth and the Shared Strokes) can all be found in older texts.[7] But this is not to say that the romance unfolds in a timeless universe. First, the poet has made a genuine attempt to fit his story into a received historical framework, ultimately derived from Geoffrey of Monmouth's authoritative *Historia regum Britanniae*.[8] The story is situated in the time of 'kynge Vtere' (1), otherwise known as Utherpendragon, King Arthur's father. For a contemporary audience familiar with Galfridian 'history', this was a well-known period of internecine strife before King Arthur subdued the enemies of the English nation in Britain and abroad.[9] In chivalric romance this 'historical' setting had already been used in Chrétien de Troyes's *Conte du Graal*, where the misfortunes of Perceval's father take place before the time of King Arthur, under whom such things would not happen:

> Sa granz terre, ses granz tresors,
> qui il avoit come prodom,
> ala tot a perdicion,
> si cheï an grant povreté.
> Apovri et deserité
> et essillié furent a tort
> li prodome aprés la mort

[7] On the motif of the Spendthrift Knight see G. H. Gerould, *The Grateful Dead* (London, 1908); on the other motifs see C. G. Loomis, '*Sir Cleges* and Unseasonable Growth in Hagiology', *Modern Language Notes* 53 (1938), 591–4; S. T. Carr, 'The Middle English Nativity Cherry Tree: The Dissemination of a Popular Motif', *Modern Language Quarterly* 36 (1975), 133–47; and J. R. Reinhard, 'Strokes Shared', *Journal of American Folklore* 36 (1928), 380–400.

[8] I discuss this technique of accommodating stories within an accepted historical framework in 'Finding Time for Romance: Mediaeval Arthurian Literary History', *Medium Aevum* 53 (1994), 1–16. The pervasiveness of these attempts to 'historicize' medieval romance is one reason why I cannot agree with Gurevich's generalization that 'Lancelot, Perceval, and other heroes of this type are ageless beings moving in a biographical vacuum' (*Categories*, p. 124).

[9] Utherpendragon's turbulent reign is recounted in part six of Geoffrey's *Historia*. See *Geoffrey of Monmouth: The History of the Kings of Britain*, trans. L. Thorpe (Harmondsworth, 1966), pp. 202–12.

> Utherpandragon, qui rois fu
> et peres le bon roi Artu. (ll. 436–44)[10]
>
> (His great land and his great wealth which he possessed as a worthy man all went to ruin, and he fell into utter poverty. After the death of Utherpendragon, the father of good King Arthur, noble men were reduced to poverty and disinherited and wrongly exiled.)

As in Chrétien's *Conte du Graal*, the turbulent period before Arthur provides an 'authentic' backdrop to Cleges's troubled life and times, when gentlemen 'traveld . . . in londe of ware / And wern fallyn in poverté bare' (16–17). To these victims of historical circumstance Cleges offers his support until in the end he, too, has nothing left to live on.

The story of *Cleges* can indeed be 'dated' even more precisely to the earliest years of Uther's reign, since an aside informs us that the cherries that Cleges gave to Utherpendragon were afterwards presented by the king to his future bride:

> The kynge therfore made a presente
> And sent yt to a lady gente
> Was born in Corneweyle.
> Sche was a lady bryght and schen;
> After sche was hys owne Quen,
> Wythouten any feyle. (ll. 385–90)

There was no need for the poet to name the lady, since his audience would have known their history: Uther married Ygerne, widow of duke Gorlois of Cornwall, having already engendered upon her a son who would be called Arthur. By acting as the go-between in the courtship of Uther and Ygerne, Cleges's cherries finally become part of British history.

The story of *Cleges* is thus carefully placed in a specific period of British 'history'. Equally purposeful is the choice of season: Christmas. *Cleges* shares this seasonal setting with a number of other Middle English romances – *Sir Perceval of Galles*, *Rauf Coilyear*, *Sir Gawain and the Green Knight*, *Amis and Amiloun*, and *The Alliterative Morte Arthure* – in all of which the story or a crucial part of it is set at Christmas. The remarkable thing about all these Christmases is that they all replace a springtime setting in the source.[11] They

[10] Chrétien de Troyes, *Le Conte du Graal (Perceval)*, ed. F. Lecoy, 2 vols (Paris, 1984). The 'historicity' of this setting is discussed by S. Sturm-Maddox, '"Tenir sa terre en pais": Social Order in the *Brut* and the *Conte del Graal*', *Studies in Philology*, 81 (1984), 28–41.

[11] In *Sir Perceval of Gales*, Christmas replaces spring in Chrétien's *Conte du Graal*. As argued by H. M. Smyser, '*The Taill of Rauf Coilyear* and Its Sources', *Harvard Studies and Notes in Philology and Literature* 14 (1932), 135–50, *Rauf Coilyear* is probably based on an earlier version of the romance of *King Edward and the Shepherd* (*Middle English Romances*, ed. W. H. French and C. B. Hale, 2 vols. [New York, 1964], II, 949–85) which is set 'one mornyng of May' (l. 15). In the probable source for the Beheading Game in *Sir Gawain and the Green*

are not, then, an inertly inherited convention but the result of changes that Middle English poets consistently made to their sources. This pattern of adaptation, which also transformed the French springtime dance-songs (*caroles*) of the High Middle Ages into the later Middle English Christmas carol,[12] may partly reflect the development of Christmas into the most popular religious feast of the fourteenth century.[13] But no doubt it also reflects the opportunism of English composers and performers, who, given a captive audience on this traditional season of liberality, must have had higher hopes than usual of a generous return for their efforts. It is surely no coincidence that medieval minstrels and singers received by convention a *largesse* at Christmas,[14] and that at Christmas gifts (including poems) were presented to others, obviously in the hope that they would be reciprocated.[15]

In *Cleges* the longed-for remuneration for poetic labour is pointedly inscribed into the story, where wish becomes reality by dint of Cleges's singular generosity to minstrels:

> Mynstrellys, when the fest was don,
> Schuld not withouten gyftes gon,
> That were both rych and gode,
> Hors and robys and rych ryngis,
> Gold and sylver and other thyngis,
> To mend wyth her mode. (ll. 49–56)

How exactly *Schuld* is to be understood in this passage is unclear. The context would favour the rare 'consuetudinal shall' ('minstrels would not go without gifts'), while the probabilities of Middle English usage favour 'hortatory shall' ('minstrels should not go without gifts').[16] The ambiguity as to whether

Knight, a version of the *First Continuation of Chrêtien's 'Perceval'*, both stages of the game take place 'Au moys de May': see 'The Prose Redaction of the *Caradoc* Beheading Game', in L. D. Benson, *Art and Tradition in 'Sir Gawain and the Green Knight'* (New Brunswick, NJ, 1965), pp. 249–57 (p. 250). The poet of the *Alliterative Morte Arthure* changes King Arthur's Whitsuntide feast of the Galfridian tradition into a Christmas feast. The Christmas setting in *Amis and Amiloun* is discussed below.

12 See *The Early English Carols*, ed. R. L. Greene, rev. edn (Oxford, 1977). In all the examples of French *caroles* given by Greene (p. xlvi) we are dealing with springtime dance-songs; in the English tradition, however, the vast majority of carols are associated with the twelve days of Christmas (p. xxxviii).

13 As suggested in *Sir Gawain and the Green Knight*, ed. T. Silverstein (Chicago, 1974), p. 115.

14 R. F. Green, *Poets and Princepleasers: Literature and the Court in the Late Middle Ages* (Toronto, 1980), p. 24.

15 Particularly revealing is the definition of 'offering' in the *Promptorium Parvulorum*, ed. A. C. Mayhew, EETS ES 102 (London, 1908), pp. 362–3: '*offerynge*: a presaunte to a lorde at Crystemasse or other tymes'. On the giving of poems on New Year's Day – particularly relevant to *Sir Gawain and the Green Knight* – see J. Brand, *Observations on Popular Antiquities*, 2 vols. (London, 1813), I, 13–14.

16 For some rare instances of 'consuetudinal shall' in Middle English see T. F. Mustanoja, *A Middle English Syntax* (Helsinki, 1960), pp. 599–600.

the poet is concerned with what *did* happen to minstrels or what *should* happen to them betrays the double agenda of this 'minstrel digression' in its oscillation between constative and performative language: the poet dwells on the customary generosity to minstrels in the past to encourage *largesse* in the present.

Significantly, minstrels feature again in the denouement of the story in the Advocates version, for by a neat dispensation of poetic justice Cleges is finally recognized (both financially and personally) when a *harpor* sings before the king and his friends the story of Cleges and thus reminds the king of Cleges's merits:

> An harpor sange a gest be mowth,
> Of a knyght there be sowth,
>   Himselffe werament.
> Than seyd the kynge to þe harpor,
> 'Where ys knyght Cleges, tell me here;
>   For thou hast wyde iwent.
> Tell me trewth, yf thou can,
> Knoweste thou of þat man?'
>   The harpor sayd, 'Yee iwysse:
>
> Sum tyme forsoth i hym knewe;
> He ys a knyght of youris full trewe
>   And comly of gesture.
> We mynstrellys mysse hym sekerly . . .' (ll. 484–96)[17]

In other words, generosity to minstrels pays, since givers gain the good will of the media and are fondly remembered in the story of *Cleges* and the 'gest' of *Cleges* within it. In the absence of any hard evidence to prove (or disprove) that *Cleges* and other 'Christmas romances' were intended to be recited at a Christmas feast, one good reason for assuming that they were is that it puts us in a position to appreciate these playfully reflexive moments in which the real world merges with the fictional world, as when an audience listening to a recital of *Cleges* on Christmas Day hears of an audience listening to a recital of *Cleges* on Christmas Day, or when an audience, listening to *Sir Gawain and the Green Knight* on New Year's Day, hears of Arthur's desire for a tale of adventure before he will eat his New Year's dinner.

The 'historical' and seasonal settings of *Cleges* show that the poet handles the broad chronological background to his story intelligently. I believe that the poet's precise chronology is similarly intelligent and would like to challenge the view that he made a mess of the calendar by leaving a whole day unaccounted for. I shall first describe the chronological problem, then

[17] I cite the useful parallel-text edition by A. Treichel, '*Sir Cleges*: eine mittelenglische Romanze', *Englische Studien* 22 (1896), 345–88. In Ashmole 61 this episode is missing, though here, too, it is a minstrel who informs the king about Cleges.

propose a solution, and finally provide some contextual support for that solution.

Cleges's temporal markers allow us to infer a reasonably precise chronology. After a brief description of Cleges's spending habits on previous occasions, the story proper begins on Christmas Eve:

> It fell on a Crystenmes Eve,
> Syre Cleges and his wyfe,
> They duelled by Cardyff syde.
> When it drew towerd the none,
> Syre Cleges fell in swownyng sone;
> Wo bethought hym that tyde,
> When myrth he was wonte to hold,
> And he, he had hys maners solde,
> Tendandrys and landes wyde. (ll. 85–93)

At this moment, around noon on Christmas Eve, Cleges suddenly hears all around him the sounds of festive music, of singing and dancing, of instruments playing (ll. 97–105). Only Cleges appears to hear these joyous sounds, and the poet delicately leaves open the question of whether they are supernatural or natural, external or internal to Cleges's 'swoon'. Thus the timing of Cleges's swoon – 'it drew towerd the none' – could be explained both psychologically (noon is mealtime and so reminds Cleges painfully of his present hardship and past 'myrth') and supernaturally, for by a tradition stretching back at least as far as the Old Testament, noon is the time when magic is most likely to strike.[18] In Middle English romance the *locus classicus* is *Sir Orfeo*, in which the Fairy King appears to the sleeping Heurodis at noon ('in an undrentide').[19] Now noon (and 'undern') could mean various things in Middle English (9am, 12am and 3pm), and, at least in *Orfeo*, there seems to be some confusion about which of these times the poet had in mind.[20] However, the rhythms of the household in *Cleges* offer us a clue. After reminding her husband that no-one should be sad 'Agene this holy day' ('as the holy day approaches'), Cleges and his wife proceed to dinner:

> Than they wesch and went to mete
> With such god as they myght gete
> And made mery chere. (ll. 154–7)

After dinner, Cleges and Clarys 'drof the dey away' (l. 158) and 'after evensong went to bede / At certeyn of the nyght' (ll. 161–2).[21] So far the

[18] See J. B. Friedman, 'Eurydice, Heurodis, and the Noon-Tide Demon', *Speculum* 41 (1966), 22–9. The title alludes to the *daemonio meridiano* of Psalm 90.6 in the Vulgate.

[19] 'Sir Orfeo', in *Middle English Breton Lays*, ed. Laskaya and Salisbury, l. 65.

[20] Laskaya and Salisbury note the different meanings of *undrentide* (*Middle English Breton Lays*, p. 47), but then mistakenly gloss the word as 'morning' at l. 133.

[21] The Advocates MS has 'after soper' (l. 161) instead of 'after evensong'. The manuscript

chronology must have been immediately recognizable to a medieval audience. The same pattern is found in *Sir Gawain and the Green Knight*: Gawain arrives at Castle Hautdesert in the late morning and is seated near a fireplace to warm up while the table is laid for a fish dinner (Christmas Eve is a fast day). After dinner, when daylight is fading (ll. 928–9), the bells ring 'To þe hersum euensong of þe hyȝe tyde' (ll. 928–9).

What timetable are we to imagine for these events? Tolkien and Gordon's gloss on *diner* in their edition of *Gawain* offers us poor guidance – 'dinner (the chief meal of the day) [was] begun by about 2 o'clock' (p. 176) – for in fact dinner was at least a couple of hours earlier. For example, the household regulations of Cicely, duchess of York (*c.* 1470),[22] which are not untypical of the period, tell us that she rose at 7.00; had her dinner on 'eating days' at 11.00 and on fast days at 12.00; went to evensong sometime between 3.00 and 5.00 in the afternoon; ate supper at 5.00; and went to bed at 8.00. The reason for postponing dinner to 12.00 is that, by the fourteenth century, dinner was normally the last proper meal to be served on fast days.[23] It seems likely that both Cleges and Gawain have their dinner at noon, and that noon and probably 'undern' in *Sir Orfeo* mean midday, around 12.00.

Cleges and his family, who have gone to bed sometime after evensong, get up the next day when the church bells ring to announce mass (l. 163). When the service is done, the family return home and Cleges wanders disconsolately in his garden, where he prays to God and discovers that the cherry tree bears fruit. Clarys's advice is to present the cherries to the king:

> 'Tomorrow, when the dey do spryng,
> Ye shall to Cardyff to the Kyng,
> Full feyre hym to presente'. (ll. 232–4)

And so it happens that, 'The morne, when it was dey lyght' (l. 241), Cleges fills his basket with cherries and travels with his son to court:

> Syr Cleges and his son gent
> The ryght wey to Cardyfe went
> On Crystenmes Dey. (ll. 253–5)

On Christmas Day? The editors are baffled: 'There seems to be an error in chronology. If the miracle occurs on Christmas Day then Cleges' journey to

reading is unlikely to be authentic: supper was not usually taken on the vigil of Christmas Day (see n. 23), and scribal corruption has left the relevant stanza (ll. 157–65) three lines short.

[22] See T. Wright, *A History of Domestic Manners in England During the Middle Ages* (London, 1862), pp. 424–5. For this and comparable medieval timetables see C. M. Woolgar, *The Great Household in Late Medieval England* (New Haven, 1999), p. 85.

[23] Woolgar, *Great Household*, p. 90. In *Gawain* the company have wine and spiced cakes before bed (ll. 979–80), but the 'soper' that is eaten on other days (ll. 1400, 1654) is omitted on Christmas Eve.

deliver the gift to King Uther takes place on Boxing Day, the day after Christmas.'[24] In short, the problem is that we have seen Cleges go to bed on Christmas Eve, get up on Christmas Day, leave the next day ('tomorrow'), and yet arrive on Christmas Day. How can this be?

The answer is really quite simple: the bells that summon Cleges and his family to church ring not in the morning but at midnight; and when Clarys tells Cleges to present the fruit 'tomorrow' she means not the next calendar day but 'on the morrow': that is, in the morning. None of this is spelt out by the poet because it would have been obvious to a contemporary audience. And I am afraid that if we are not clear about the poet's timetable we have misunderstood the best part of the poem. What the poet and his audience knew was that Christmas Day had three masses, the first at midnight (*missa in gallicantu* or *missa in nocte*), the second at dawn (*missa in aurora*) and the third at midday (*magna missa*).[25] This at least was the liturgical theory; whether in practice ordinary laymen attended all three masses is another question. From a precious early sixteenth century notebook of a Bavarian priest (for which there exists no Middle English equivalent),[26] it would appear that the second of these three masses could be omitted, but not the first and the third, during which the church was packed: the third because it was the most solemn of the three masses, and the first because midnight was the time that Christ was born. As the collect of the Midnight Mass recounts: 'Fulserunt et immania *nocte media* pastoribus lumina' ('And *at midnight* a great light shone upon the shepherds').[27]

That in fourteenth-century England, too, the laity turned out *en masse* for the midnight service is clear from Robert Mannyng's *Handlyng Synne*,[28] in which Mannyng retells the exemplary story of the riotous carollers, adding many details presumably based on his own experience. Like *Cleges* this story begins 'vpp-on a crystemesse nyȝt' (l. 9015) as the priest is saying 'matynes' (l. 9055). (On Christmas Day Matins was followed immediately by the Midnight Mass.)[29] Disturbed by the noise of rowdy revellers, the priest admonishes them from the porch:

[24] *Middle English Breton Lays*, ed. Laskaya and Salisbury, p. 403.

[25] On the medieval feast of Christmas see C. A. Miles, *Christmas in Ritual and Tradition* (London, 1912). A useful description of the liturgy of Christmas Day is provided by A. Adam, *The Liturgical Year*, trans. M. J. O'Connell (New York, 1982), pp. 212–30. On the medieval terms for the three Christmas masses, see A. Hughes, *Medieval Manuscripts for Mass and Office: A Guide to Their Organization and Terminology* (Toronto, 1982), pp. 124–5.

[26] E. Bishop, 'Pastor Dreygerwolt's Note-Book', in his *Liturgica Historica: Papers on the History and Religious Life of the Western Church* (Oxford, 1918), pp. 444–52.

[27] The Latin text is cited from *Missale ad Usum Insignis et Praeclarae Ecclesiae Sarum*, ed. F. H. Dickinson (Burntisland, 1861–83), p. 50. The English translation is from *The Sarum Missal in English*, trans. F. E. Warren, 2 vols. (London, 1911), I, 96.

[28] *Robert of Brunne's 'Handlyng Synne'*, ed. F. J. Furnivall, 2 vols., EETS OS 119, 123 (London, 1901; repr. 1997).

[29] K. Young, *The Drama of the Medieval Church*, 2 vols. (Oxford, 1933), I, 547.

'But comyth yn, on feyre manere,
Goddes servuyse for to here,
And doþ at Crystyn mennes lawe . . .
Wurschyppeþ hym with alle ȝoure myȝt,
þat of þe vyrgyne was bore þys nyȝt.' (ll. 9067–72)

Attendance at Midnight Mass was obviously expected as the 'lawe' of Christian men.

A technical objection to my proposed timetable for *Cleges* would be the poet's use of 'tomorrow' (l. 232) and 'the morne' (l. 241), which, if I am right, would have to refer, not to the subsequent calendar day, but to the approaching daytime.[30] But this is only to be expected: if we say 'tomorrow' in the early hours of the morning we usually have in mind the next natural day rather than the next calendar day. The context-dependence of 'morrow' and the customary attendance at Midnight Mass are well illustrated by *Amis and Amiloun*.[31] In this early fourteenth century romance, Amiloun, who has been afflicted with leprosy, finds shelter with his blood-brother Amis. Both friends are visited in their sleep by an angel who announces that Amiloun can be cured if he is washed in the blood of Amis's children at the same time as 'Jhesu Crist was born' (l. 2204). As the fateful day approaches, Amis's wife and the household servants leave the castle to attend mass:

So it bifel on Cristes night,
Swiche time as Jhesu, ful of might,
  Was born to save mankunne,
To chirche to wende al that ther wes,
Thai dighten hem, withouten les,
  With joie and worldes winne. (ll. 2251–6)

Only Amis stays behind, saying that he wishes to keep his friend Amiloun company. Alone in the house, he takes a 'candel fair and bright' and enters the room where his children lie asleep, kills both, and bathes Amiloun in a basin filled with their blood.

The Christmas-night setting is unique to the Middle English version: in the probable Anglo-Norman source the killing of the children takes place 'one day'.[32] And yet, despite being a late addition to the story, the setting has been fully realized by the poet, and it touches with admirable sureness the story's emotional core. For the night that 'Dryȝtyn for oure destyné to deȝe watz

[30] But cf. 'And on the morowe they went to chirch' (Advocates MS, 163), where 'on the morowe' means 'the next day'.

[31] 'Amis and Amiloun', in *Of Love and Chivalry: An Anthology of Middle English Romance*, ed. J. Fellows (London, 1993).

[32] *Amys et Amillyoun*, ed. H. Fukui, Anglo-Norman Text Society, Plain Text Series 7 (London, 1990), l. 1084. In the French *chanson de geste*, Amis murders his children one Sunday morning, when everyone has gone to mass: *Ami et Amile: chanson de geste*, ed. P. F. Dembowski, Classiques français du Moyen Age (Paris, 1987), ll. 2796–9.

borne' (*Gawain*, l. 996) not only offers Amis the opportunity to carry out the murder unhindered, in a dark and empty house, but also helps the poet to emphasize the redemptive significance of the sacrifice Amis is asked to make. While everyone in church celebrates Christmas – which in the medieval period meant celebrating the prospect of Redemption – Amis redeems Amiloun by sacrificing his own children. The point of the synchrony of the sacrifice and the mass is to expose the child-murder as Amis's private religious ritual, which commemorates *in deed* what the congregation would have heard *in word*, in the reading from the epistles at Midnight Mass (1 Titus 2.11–15):[33] the Christ-child 'gave himself for us, that he might redeem us from all iniquity.'

In the early morning Amis's wife returns with the other members of the household to find Amiloun fully recovered and the children miraculously restored to life:

> Amorwe, as tite as it was day,
> The levedi com hom, al with play,
> With knightes ten and five. (ll. 2353–4)

And that *morwe*, to avoid any confusion, is the morning of Christmas Day and not the day thereafter. Looking closely at the immediate context of these Christmas 'morrows' in *Amis and Amiloun* and *Cleges*, I am struck by the care taken by both poets to disambiguate the implied chronology. Compare *Amis and Amiloun* l. 2352 ('Amorwe, as tite as it was day') with *Cleges* l. 232 ('Tomorrow, when the dey do spryng') and l. 241 ('The morne, when it was dey lyght'). These sub-clauses are not empty line-fillers but instruct us move our imaginary clocks forward from midnight to the morning of Christmas Day and not from one day to the next.

A number of other details in *Cleges* also fall into place once we set our mental clocks after the poet's. We understand, for example, why Cleges tells the usher '*Thys nyght* this fruyt grew' (l. 307), and why Cleges has to discover the fruit in such a roundabout way:

> As he knelyd oune his kne
> Underneth a chery tre,
> Makyng a praere,
> He rawght a bowghe in hys hond,
> To ryse therby and upstond;
> No lenger knelyd he ther.
> When the bowghe was in hys hond,
> Gren levys theron he fond
> And ronde beryes in fere. (ll. 193–201)

On the one hand, the haphazardness of Cleges's discovery is suggestive of

[33] See above, n. 27.

forces outside human control, of the necessary intervention of the divine grace that answers Cleges's prayer, but Cleges's fumbling also shows how fully the poet has imagined this moment of midnight darkness: since Cleges cannot *see* the leaves and the cherries of the tree in the darkness of his garden, he must *feel* them.

And it is also the poet's sense of time that distinguishes his deployment of the motif of the miraculous cherries in *Cleges* from that in other later Middle English texts, in which it also occurs.[34] These texts are the play of *Joseph and the Midwives* from the N-Town Cycle and the 'Cherry Tree Carol'. In both texts the miracle occurs when Joseph and Mary are on their way to Bethlehem, and Mary, seeing some ripe cherries, asks Joseph to pluck them for her. Unable to reach them, Joseph quips sarcastically that she should ask the man who got her pregnant – upon which the tree miraculously bends down to present its fruit to Mary. In the 'Cherry Tree Carol' the unborn Christ orders the tree to bow down to Mary. Sherwyn T. Carr has argued that the *Cleges*-poet derived the motif from the N-Town play (which also mentions their miraculous 'unseasonableness'), but the suggestion is implausible both on grounds of date and comparability.[35] For what is most significant about the motif in *Cleges* is that the poet has carefully synchronized this miracle with the liturgical time of Christ's birth. And for that idea analogues far more compelling than the N-Town Cycle exist. I am referring to the legend, first recorded in the early fifteenth century, that apple trees flowered and bore fruit on Christmas night;[36] to the legend, associated with Joseph of Arimathea, that the walking stick he planted in Glastonbury is now a rosebush that blossoms each year in honour of Christ's birth, just as the nearby hawthorns 'do burge and bere grene leauues at Christmas';[37] and finally to the thirteenth-century Life of St Hadwigis, which records the following miracle:

> Quoniam eo tempore, dum adhuc iuvenis esset, in dei natalis domini venit qui diceret coram ea sedente in mensa, quod arbor una cerasus stans in horto recentibus esset floribus decorata.[38]
>
> (Once at that time, when she was still young, somebody came on Christmas Day saying while she was seated at the table that a cherry tree in the garden stood in full flower.)

[34] S. T. Carr discusses the origin of the motif in the Pseudo-Matthew gospel and its attestations in Middle English texts. See n. 7.

[35] The motif of the cherry-tree was added to the play of *Joseph and the Midwives* by a late reviser, sometime in the 1460s. If the play were a source for *Cleges*, the latter would have to be 'a late-fifteenth-century narrative poem', as Carr indeed claims (p. 133).

[36] See A. Tille, *Yule and Christmas: Their Place in the Germanic Year* (London, 1899), pp. 170–3.

[37] 'The Verse "Lyfe of Joseph of Armathia"', in *The Alliterative Romance of Joseph of Arimathie*, ed. W. W. Skeat, EETS OS 44 (London, 1871), l. 386.

[38] Cited from Tille, *Yule*, p. 171.

Of course, none of these legends need to be direct sources. What makes them pertinent to *Cleges* is that they reveal the same current of thought that runs through *Cleges* and has its source in the medieval apprehension of the nativity as just such a 'miraculous' fruition. As St Ambrose had written in his famous hymn *Veni, redemptor gentium* (sung at Vespers on Christmas Eve):

> Verbum Dei factum caro
> Fructusque ventris floruit. (ll. 7–8)[39]
>
> (The word of God is made flesh and the womb's fruit has flourished.)

But why cherries? Numerous symbolic meanings might be adduced,[40] but none as important as the simple fact – remembered in a number of medieval proverbs – that cherries have a particularly short season.[41] Cherries are the right choice, better than apples or rosebushes, for a poet wishing to heighten the 'impossibility' of this miracle, which depends for its effect on timing alone: on the season of Christmas and on the precise liturgical hour of Christ's birth, when one miraculous midnight fruit inevitably commemorates another.

SIR GAWAIN AND THE GREEN KNIGHT

I turn now to *Sir Gawain and the Green Knight*, which presents a problem of chronology that, I am afraid, will prove more tenacious. Again I shall first outline the problem, then provide some cultural context for it, and finally consider some solutions. On his way to the Green Chapel, where he has promised to receive the Green Knight's blow on New Year's Day (which was, by the fourteenth century, 1 January), Gawain spends Christmas in Castle Hautdesert. As we have seen, he arrives there on Christmas Eve and spends Christmas Day convivially – after which the poet summarizes the days that follow:

> Much dut watz þer dryuen þat day and þat oþer,
> And þe þryd as þro þronge in þerafter;

[39] *Analecta hymnica medii aevi*, ed. G. M. Dreves and C. Blume, 55 vols. (Leipzig, 1886–1922), II, 36. Lines from this hymn are incorporated in a number of Middle English carols: see *Early English Carols*, ed. Greene, p. lxxxviii. For one of many expressions of this idea in the vernacular, see the Middle English lyric 'Upon my ryght syde y me leye', in *Late Medieval English Lyrics and Carols*, ed. T. G. Duncan (Harmondsworth, 2000), p. 106: 'Ouyre Lorde is the frwte, oure lady is the tree – / Blessed be the blossome that sprange of the' (ll. 8–9).

[40] See G. Ferguson, *Signs and Symbols in Christian Art* (New York, 1954), p. 29.

[41] See D. Gray's note on 'Farewell this world', l. 8 ('This lyfe, I see, is but a cheyrye fayre'), in his *Selection of Religious Lyrics* (Oxford, 1975), p. 154.

> þe ioye of sayn Jonez day watz gentyle to here,
> And watz þe last of þe layk, leudez þer þoȝten.
> þer were gestes to go vpon þe gray morne . . . (ll. 1020–4)
>
> (Much joy was made that day and the next, and after that the third pushed its way in as eagerly; the joy of St John's Day was delightful to hear and was the last of the party, people there thought. There were guests who had to go on the grey morning.)

In these lines a number of days pass: Christmas Day and 'þat oþer' (that is, 26 December, St Stephen's Day); the third day that 'þronge in þerafter' is St John's Day (27 December), which is the last of the Christmas 'layk', since the guests will be leaving the following morning (28 December, Holy Innocents' Day).

Or so the passage says. For the rest of the romance proceeds as if another day had passed and we had been fast forwarded to 28 December, with the guests leaving on the 29th (St Thomas Becket's Day). The contradiction emerges only forty lines later when Gawain refuses the lord's invitation to stay longer, explaining how little time remains before his appointment at the Green Chapel: 'Naf I now to busy bot bare þre dayez' (l. 1066). If Gawain has only three days left to find the Chapel where he is due on 1 January, it must be the end of 28 December. This dating is confirmed by the remainder of Gawain's stay. Still on the evening of 28 December, Gawain agrees to prolong his stay when his host tells him that the Green Chapel is only two miles away, and guest and host promise to exchange winnings on three consecutive days. Next morning (29 December) when the guests have left, the lord goes out hunting, Gawain is visited in his bedroom by the Lady of the Castle, and the first winnings are exchanged. The same happens the following day (30 December), and that evening Gawain and the lord look forward to exchanging winnings for the last time on 'Nwe ȝeres euen' (l. 1669). This next day (31 December) sees the third and last of the temptation scenes and exchanges of winnings. The morning thereafter (1 January) Gawain departs to face the Green Knight.

It would thus appear that in the poet's summary (ll. 1020–3) a whole day has gone missing. As Tolkien and Gordon note in their edition: '28 December is not accounted for' (p. 104). What is to be done? One possibility is to do nothing at all, and to dismiss this counting of days as academic pedantry that is inappropriate to what Gurevich terms the 'mytho-poetic time' of medieval romance.[42] But the *Gawain*-poet's ticking off of calendar days surely indicates that he is not in 'mytho-poetic' time. The missing day really does matter, and I think that only a correct calendar (25–28 December) can make comprehensible otherwise puzzling behaviour in the poem. To begin with Gawain's

[42] Gurevich, *Categories*, p. 124.

behaviour. As we have seen, he refuses his host's offer to stay longer on the grounds that time is running out. But if he genuinely felt pressed for time then why did not he take his leave a day or two earlier? And then there are the guests. Why do they all leave at the same time, on the exact same morning that Gawain initially intends to go? It looks very much as if everyone in the poem obeys a temporal rule that the poet and his audience take for granted – the 'layk' ends on 28 December – but which no-one, as far as I am aware, has thought it possible (or necessary) to explain.

But perhaps an explanation can be given. While Christmas was generally regarded as a twelve-day festival ending on Epiphany, a smaller number of Christmas days were classed as feast-days in the strict sense: that is, as days on which laymen (excepting kitchen and chamber staff) had to abstain from all secular occupations, including unnecessary travel, and attend church services.[43] The Latin term for such days was *festa ferianda* (as distinct from the hundreds of other *festa* that did not need to be observed by ordinary laymen).[44] The calendar of *festa ferianda* is perhaps relevant to *Gawain*. I quote the calendar, ultimately based on Gratian's *Decretum* (De cons. D. 3, c. I), as it appears in Thomas Chobham's thirteenth-century *Summa confessorum*:

> Feriandi vero per annum isti sunt dies: natalis domini, sancti Stephani, sancti Ioannis euangeliste, Innocencium, sancti Silvestri, Octave Natalis domini, theophanie . . .[45]
>
> (The following days are to be celebrated every year: Christmas Day, St Stephen's, St John the Evangelist's, the Innocents', St Silvester's [31 December], the Octave of Christmas Day [1 January], Epiphany . . .)

To these official holidays others might be added by bishops, which is why rules for the observance of feast-days tended to differ from diocese to diocese. In many English dioceses, the optional memorial of St Thomas Becket (29 December) was included amongst the *festa ferianda*. This was the case, for example, in the diocese of Hereford, where a mid fourteenth century statute specifies as holidays 'Dies natalis domini cum iiii^{or} diebus sequentibus'.[46] However, the minimum sequence as stipulated in Gratian's *Decretum* consisted of Christmas Day and the three days of St Stephen, St John and the Holy Innocents. These saints – whose martyrdom was celebrated immediately after Christ's Nativity – were known in the Middle Ages as the

[43] See E. C. Rogders, *Discussion of Holidays in the Later Middle Ages* (New York, 1940). On the subject of travel, Rodgers refers to the Middle English 'Dives and Pauper', and to Antonio of Florence, who believed that 'one should forego a day's journey unless force of circumstance or the performance of some pious task impelled one to take the road' (p. 42).

[44] See C. R. Cheney, 'Rules for the Observance of Feast-Days in Medieval England', *Bulletin of the Institute of Historical Research* 34 (1961), 117–47.

[45] T. Chobham, *Summa confessorum*, ed. F. Broomfield (Louvain, 1968), p. 266.

[46] Cheney, 'Rules', p. 144.

'companions of Christ' (*comites Christi*),[47] because in the liturgy they accompanied the Christ child in a 'cortege of honor'.[48] A mid fourteenth century list of mandatory feast-days from the cartulary of Faringdon (Berkshire, diocese of Salisbury) accordingly reads:

> Quibus diebus non laborant conversi et famuli in manerio de Farendon: In Natali domini et tribus diebus sequentibus.[49]
>
> (As regards the days on which the lay-brethren and servants at the manor of Faringdon do not work: on Christmas Day and three subsequent days.)

The special status of the three feast-days following Christmas Day is also evident from a mandate from Bishop John de Grandisson (dated 7 January 1360) to the clerics of Exeter Cathedral, warning them, under pain of excommunication, to behave themselves in a dignified manner. It had come to the bishop's attention that his clerics had amused themselves with unseemly games,

> in sanctissimis Dominice Nativitatis, ac Sanctorum Stephani, Johannis, Apostoli et Euangeliste, ac Innocencium Sollepmniis, quando omnes Christi Fideles Divinis laudibus et Officiis Ecclesiasticis devocius ac quiescius insistere tenentur . . .[50]
>
> (during the holiest solemnities of Christmas Day, and those of the feasts of St Stephen, St John the Apostle and Evangelist, and the Innocents, when all faithful Christians are obliged to devote themselves with even greater piety and humility to praising God and to church services . . .)

This fixed sequence of Christmas day followed by that of the three *comites Christi* could explain why Gawain will not leave earlier: he dutifully observes the prescribed feast-days. When the *festa ferianda* come to an end, he prepares to leave, as do the guests for whom it will be business as usual on 29 December (at least if their manors were at all like Faringdon's).[51]

Turning back now to the problematic lines of *Gawain* (ll. 1020–3) we can learn two things: firstly, by 'þe layk', the 'playtime', the poet and his audience presumably understood the official days of vacation; and, secondly, the

[47] Gulielmus Durandus, *Rationale divinorum officiorum* (Naples, 1859), VII, 42, 1.

[48] Adam, *Liturgical Year*, pp. 139–40.

[49] Cheney, 'Rules', p. 142.

[50] *The Register of John de Grandisson*, ed. F. C. Hingeston-Randolph, 3 vols. (London, 1894–9), III, 1213.

[51] It is interesting to note that the *Gawain*-poet also observes this distinction between an obligatory and an optional feast day for All Saints' and All Souls' (1 and 2 November respectively). Gawain spends the 'fest' of 'Al Hal Day' (All Saints') at Arthur's court and leaves early on All Souls' Day (ll. 536–7, 566). The choice of All Souls' for Gawain's departure has been interpreted symbolically (Gawain is doomed to become one of the faithful dead commemorated on that day). But the everyday reality of medieval holy days again seems relevant: the feast of All Saints' had to be observed by ordinary laymen, whereas All Souls' did not.

problem of the missing day has only got worse, since the omission of Holy Innocents' Day offends not only against modern expectations of verisimilitude but also against medieval expectations of Christmas and canon law. I find it inconceivable that the poet could have 'forgotten' this official feast-day.

The context of the *festa ferianda* thus supports the supposition that the text of *Gawain* is corrupt and needs emending. This course of action was first adopted by Sir Israel Gollancz, who after line 1022 added a fine alliterative line of his own invention: 'With moste myrþe and mynstrelsy Childermas sued'.[52] Modern editors accept Gollancz's solution: Theodore Silverstein prints Gollancz's line as 1022a, while Tolkien and Gordon and Andrew and Waldron canvass the hypothesis that a line has dropped out in a note to the otherwise unemended text.[53] In this they are almost certainly mistaken, for the extra line spoils the poet's numerical design whereby the first alliterative long line – 'Siþen þe sege and þe assaut watz sesed at Troye' – is repeated in the last: 'After þe segge and þe asaute watz sesed at Troye.' The number of that last alliterative long line is 2525, and thus recalls the pentangle, whose number suits Gawain since he was 'always faithful in five respects and five times under each' (l. 632). While I would not subscribe to the esoteric meanings that some critics have attached to the poet's numerology,[54] I do agree with them that line 2525 does not have that number by accident. For if it were accidental, then why does *Pearl* – which contains a vision of the heavenly Jerusalem where everything comes in twelves – have 1212 lines, the first repeating the last? No, the lines of *Gawain* are a *numerus clausus* and no day can be intercalated in the calendar by adding an extra line.

However, a far simpler hypothesis would be to assume scribal corruption not *between* the lines but *within* line 1023, which can in fact be corrected by the addition of a single adjective or adverb before *watz*, such as *lef* or *þen*.[55] With

52 *Sir Gawain and the Green Knight*, ed. Sir I. Gollancz, EETS OS 210 (London, 1938).

53 *The Poems of the 'Pearl' Manuscript*, ed. M. Andrew and R. Waldron (York, 1978).

54 For example, P. Sadowski, *The Knight on His Quest: Symbolic Patterns of Transition in 'Sir Gawain and the Green Knight'* (Newark, NJ, 1996), pp. 116–19; and B. Kowalik, *From Circle to Pentangle: Space in the Poems of the 'Pearl' Manuscript* (Lublin, 1997), pp. 73–96.

55 The least invasive emendation would be the substitution of *And* (MS &) in l. 123 by *As* (or *Als*). This deserves consideration, since the scribe of BL Cotton Nero A.x has a history of miscopying *And* (at *Patience* l. 520, *Gawain* ll. 1281, 2205), and since the error could easily be explained as mistaken anticipation of copy (compare the identical error at *Piers Plowman: The A-Version*, ed. G. Kane [London, 1960], prol. 102). However, attestations of this particular use of *as* (without preceding correlative *as*) are rare in Middle English. Chadwyck-Healey's *Literature Online* (EP version 98:1) has provided me with only one clear instance, from Gilbert Hay's *Buik of Alexander*: 'He had [twa] sonnys, war worthy men and wise, / As was þare fader . . .' (ll. 1988–9). A further difficulty is that the grammatical subject of l. 1022 is 'þe joye', which cannot really be compared with a day ('þe last of þe layk').

so many fitting words to choose from, we should perhaps be guided by paleographical considerations, and I would tentatively propose the following emendation:

> þe ioye of Sayn Jonez Day watz gentyle to here,
> And [wale] watz the last of þe layk, leudez þer þoȝten.

The syllabic and alliterative pattern of the emended line conforms to the poet's practice elsewhere;[56] and, with the addition of an adjective, the syntactic pattern follows the pattern established in the previous lines. Moreover, *wale* 'excellent, special' has this advantage over other possibilities that its omission by a scribe is readily explicable in terms of *homœoarchy*, due to its similarity with the following *watz*.

CONCLUSION

Of course, to readers concerned with time in medieval fiction, the question of how to emend will matter less than my argument that we should emend, and that the calendar clearly did matter to the *Gawain*-poet and his contemporaries – the *Cleges*-poet included. Perhaps the difficulty for the modern reader is that the action of *Gawain* and *Cleges* takes place in a dimension of time that is becoming increasingly difficult for us to apprehend. We could call this dimension 'Church time', in Jacques Le Goff's memorable phrase,[57] as long as we remember that its rhythms were matters of ordinary experience for medieval people. In *Cleges*, 'Church time' solves the problem of the missing day and explains why the poet chose to set his miracle at midnight. In *Sir Gawain and the Green Knight*, 'Church time' – more specifically the calendar of *festa ferianda* – explains what the poet meant by the 'layk' of Christmas and why Gawain does not take his leave from Bertilac before the end of Holy Innocents' Day. Although it is often argued that medieval writers were careless about matters of chronology,[58] these romances would support a very different thesis: that the impression of 'medieval indifference to time' is only the reflex of our own indifference to medieval time.

[56] On the standard analysis the emended line would be an extended alliterative line (xaa/ax) which is one of the five metrical possibilities for the a-line allowed by H. Duggan in 'Meter, Stanza, Vocabulary, Dialect', in *A Companion to the Gawain-Poet*, ed. D. S. Brewer and J. Gibson (Cambridge, 1997), pp. 221–42 (p. 224). A comparable line in terms of metre is l. 1074: 'And cum to þat merk at mydmorn to make quat yow likez'.

[57] Le Goff, *Time, Work, and Culture*, p. 33.

[58] See the references in Gurevich, *Categories*, p. 332, n. 133.

7

Meditations on the 'Historical Present' and 'Collective Memory' in Chaucer and *Sir Gawain and the Green Knight*

Mary Carruthers

> [S]peculation on time is an inconclusive rumination to which narrative activity alone can respond.[1]

> Inde mihi visum est nihil esse aliud tempus quam distentionem: sed cuius rei nescio, et mirum, si non est ipsius animi.
> (I have therefore come to the conclusion that time is nothing other than tension: but tension of what I do not know, and I would be very surprised if it is not tension of consciousness itself.)[2]

This study will address a problem which I am still trying to work through, and so I apologize in advance if I go off in various directions and if my connections do not always seem entirely apparent. I began thinking about the topic of present tense use in Middle English narratives, especially *Sir Gawain and the Green Knight*, nearly twenty years ago and when I started then my ruminations took me almost at once to the subject of memory. Memory provided sufficiently rich material that I suspended my interest in tense, while still suspecting that there was a link between the two subjects. After all, St Augustine had insisted upon one, though (as so often) rather cryptically, for instance in the passage which forms the second of my initiating quotations.

The problem I want to get at is encapsulated in that familiar oxymoron, the 'historical present'. Virtually all linguists will immediately point out that this oxymoron is not real, that the 'history' in the 'historical present' means 'story' and that this tense is thus more properly and coherently called 'the narrative present'. But this bit of academic tidiness does not dispose of my problem, because the oxymoron is in fact a real one – there is an irreducible paradox in the familiar phenomenon of a 'historical present' that switching names will not solve. I will address the problem here through examples from Middle English literature, especially from *Sir Gawain and the Green Knight*, but I mean

[1] P. Ricoeur, *Time and Narrative*, trans. K. McLaughlin, 3 vols. (Chicago 1984–8), I, 6.

[2] Augustine of Hippo, *Confessiones*, ed. L. Verheijen, CC 27 (1990), XI.26 and Augustine of Hippo, *The Confessions*, trans. M. Boulding (London, 1997), XI.26.

to show that the paradox is general enough to affect most efforts to compose not just stories but histories, and that my own way through the paradox is to place it within the context of memory-making, especially of the sort called (in another seemingly incoherent concept) 'collective memory'.

The use of the present for story-telling is a phenomenon well known in antiquity and ubiquitous still in ordinary speech. Some recognizable form of it is known in virtually all languages and cultures. Its use has ordinarily, in western rhetoric, been analysed as an aspect of style, and attributed to various mimetic motives on the part of the story-teller, such as to imitate continuous action, or to bring an event 'before the eyes' and so into the 'now' of the actual reader or audience or to foreground and emphasize particular actions over others (this latter is a variant of the 'vividness' criterion). It has been noted often, for example, that in the *Chanson de Roland* (probably the text in which the phenomenon of the historical present has been most fully studied) an event will be introduced in the completed past (the *passé simple*) but the poet will then immediately switch to the present, using it until the event is finished.[3] For example, note the tense uses in the following sentence:

> Roland *ad mis* [Past] l'olifan a sa buche,
> *Empeint* [Pres.] le ben, par grant vertut le *sunet* [Pres.] (ll. 1753–4)
> (Roland has placed the olifant at his lips, He positions it well, sounds it with all his might.)[4]

The historical present phenomenon in Chaucer was described by L. D. Benson in a 1961 essay in *English Studies*.[5] Among many others, he notes the following instance from Chaucer's 'The Shipman's Tale':

> The morwe *cam* [Past], and forth this marchant rid*eth* [Pres]
> To Flandres-ward; his prentys wel hym gyd*eth* [Pres],
> Til he *cam* [Past] into Brugges murily. (ll. 299–301)[6]

Benson characterizes such usage as typical of Chaucer: an event introduced by a past tense and ended by a past tense, but the activity itself carried by the present. The effect can also give an impression of vividness and immediacy

[3] See, esp. the important essay by A. G. Hatcher, 'Tense-usage in the *Roland*', *Studies in Philology* 39 (1942), 597–624, which established the basis of much further work on this aspect of the *Roland*. Recent work includes M.-L. Ollier, 'Le présent du recit: temporalité et roman en vers', *Langue Français* 40 (1978), 99–112; and S. Fleischman, '"Overlay Structures" in the *Song of Roland*', *Proceedings of the Berkeley Linguistics Society* 12 (1986), 108–23.

[4] References are to *Chanson de Roland*, ed. F. Whitehead, 2nd edn (Oxford, 1946).

[5] L. D. Benson, 'Chaucer's Historical Present: Its Meaning and Uses', *English Studies* 42 (1961), 65–77. See also G. Bauer, 'Historisches Präsens und Vergegenwärtigung des epischen Geschehens: ein erzhltechnischer Kunstgriff Chaucers', *Anglia* 85 (1967), 138–60.

[6] All references are to *Riverside Chaucer*.

(rhetorical *enargeia*, or 'bringing-before-the-eyes'), but tense-derived vividness is restricted by Chaucer to *events* of great activity or 'liveliness,' *energeia*.[7] So, the present is used extensively in the Knight's account of the great joust between the armies of Palamoun and Arcite, but not in his formal descriptions, such as the 'General Prologue' portraits, where 'vividness' might most be expected.

Though I think there is some truth to Benson's definition, the problem with all these efforts to characterize the choice of the present tense in medieval literature is that other instances (indeed, the majority of them in some texts) simply do not tidy up in this way. It takes a self-consciously 'grammatical' stylist to use the historical present consistently: thus, most modern writers who select it are attempting to imitate some temporal quality like 'immediacy', 'transparency', or 'colloquialness'. Sports reporting, for example, is conducted in the present tense even though the events reported have, of necessity, occurred before the reporter can describe them to us; indeed, unstructured reporting has now come to be an important aspect of present tense usage.[8] The French 'new novelists' of the sixties, like Alain Robbe-Grillet and Marguerite Duras, were self-consciously in revolt against the notion of plot when they adopted the present tense exclusively for their works. Roland Barthes claimed that the present was the tense of immediate action and momentary perception; and thus that stories, which make connections among events, and descriptions, which structure perception, are naturally related in the past because only later reflection and recollection can create such connected designs.[9] There was a time early in the twentieth

[7] Aristotle analysed a certain type of metaphor that brings a subject 'before-the-eyes' ('prò ommaton') as possessing a quality he called *energeia*, animation: Aristotle, *Ars rhetorica*, ed. R. Kassel (De Gruyter, 1976), III.11.1411b. This came to be confused in later rhetoric with *enargeia*, the stylistic quality of raising up mental images, *phantasiai*, for persuasive emotional effect (*pathos*). In Roman forensic rhetoric, the demonstrative importance of *phantasiai* (*visiones* in Latin) that bring one's subjects before-the-eyes is emphasized by Quintillion, *Institutio oratoria*, ed. M. Winterbottom, 2 vols. (Oxford, 1970), VIII.iii.61–70, who praises particularly Cicero's masterful use of *enargeia* in his courtroom speeches. It is worth noting that Cicero did not use the present tense in such moments, though he sometimes used a present passive like 'videor' to introduce a vivid description. A useful discussion of the differences in Greek between *energeia* and *enargeia* and the critical role of the animate image in human judgment is K. Eden, *Poetic and Legal Fiction in the Aristotelian Tradition* (Princeton, 1986), pp. 69–111.

[8] 'Performance' and thus orality is frequently cited as natural to the narrative use of the present. See N. Wolfson, 'A Feature of Performed Narrative: The Conversational Historical Present', *Language in Society* 7 (1978), 215–37; and W. Ong, 'Oral Remembering and Narrative Structures', in *Analyzing Discourse: Text and Talk*, ed. D. Tannen (Washington, 1982), pp. 12–24.

[9] The use of historical present in post-medieval narratives is far more self-consciously justified and rationalized than it is in medieval vernacular literatures (where it isn't at all), because by the eighteenth century all the western European vernacular literatures had been given a rationalized grammar, mostly on the model of literary Latin. Some interesting analyses of modern uses of the present in narrative include A. Banfield, 'Grammar and

century when use of the story-telling present signalled lower class and lack of education: ladies told stories properly, using the past tenses. But these regularities all result from academic grammarians imposing their logics on language: doing what a friend of mine, who often has to write profoundly, if arbitrarily, rationalized accounts of architectural projects, calls 'post-design justification'.

For the embarrassing characteristic of medieval narratives is that their tense selection follows no particular rationale. A typical vernacular narrative switches readily from past to present and back again several times within the same event, and sometimes within the same sentence or even clause. This phenomenon has been observed for as long as scholars have edited medieval texts, particularly in Old French narratives: indeed French scholarship refers to it as *confusion des temps*.[10] Most of the modern scholarship dealing with it has studied French and other Romance literatures (Spanish, Occitan), so its apparent commonness in French may be, at least partly, a consequence of modern scholarly activity. Tense switching also appears to be more frequent in verse than in prose; it is far more common in the vernaculars than in Latin, and, so it seems, in Romance vernaculars than in Germanic ones. For a time, some English scholars claimed that tense switching was rare in medieval English literature (remember the class prejudice characteristic of English speakers early in this century against the use of the story-telling present). It does occur far more often in Middle than in Old English narrative, though it is by no means unknown in Old English, as was once claimed. Here is a familiar passage from *Sir Gawain and the Green Knight*, the opening stanzas.[11] I have highlighted the tenses used:

> Siþen þe sege and þe assaut *watz sesed* [ImPast] at Troye,
> þe borȝ brittened and brent to brondez and askez,
> þe tulk þat þe trammes of tresoun þer *wroȝt* [Past]
> *Watz tried* [ImPast] for his tricherie, þe trewest on erthe:
> Hit *watz* [Past] Ennias þe athel, and his highe kynde,
> þat siþen *depreced* [Past] prouinces, and patrounes *bicome* [Past]

Memory', *Proceedings of the Berkeley Linguistics Society* 11 (1985), 387–97 (on Proust); the chapters on French *nouveau roman* in S. Fleischman, *Tense and Narrativity: From Medieval Performance to Modern Fiction* (Austin, 1990); and L. J. Brinton, 'The Historical Present in Charlotte Bronte's Novels: Some Discourse Functions', *Style* 26 (1992), 221–44. Major modern narrative theory has been based on the unmarked 'naturalness' of the past tense in narrative, perhaps most influentially argued by Roland Barthes: see, e.g., the first chapter of *Writing Degree Zero*, trans. A. Lavers and C. Smith (New York, 1967). Fleischman, *Tense and Narrativity*, pp. 94–123 has an analysis of some of the limitations of modern narrative theory in the medieval context.

10 Fleischman, *Tense and Narrativity*, pp. 7–10.

11 Two studies of the use of historical present in *Sir Gawain and the Green Knight* are R. Zimmerman, 'Verbal Syntax and Style in *Sir Gawain and the Green Knight*', *English Studies* 54 (1973), 533–43; and P. Richardson, 'Tense, Discourse, and Style: The Historical Present in *Sir Gawain and the Green Knight*', *Neuphilologische Mitteilungen* 92 (1991), 343–9.

Welneȝe of al the wele in þe west iles.
Fro riche Romulus to Rome *ricchis* [Pres] hym swyþe,
With gret bobbaunce þat burȝe he *biges* [Pres] vpon fyrst,
And *neuenes* [Pres] hit his aune nome, as hit now *hat* [Pres];
Tirius to Tuskan and teldes *bigynnes* [Pres],
Langaberde in Lumbardie *lyftes* [Pres] vp homes,
And fer ouer þe French flod Felix Brutus
On mony bonkkes ful brode Bretayn he *settez* [Pres]
with wynne,
Where werre and wrake and wonder
Bi syþez *hatz wont* [ImPast] þerinne,
And oft boþe blysse and blunder
Ful skete *hatz skyfted* [ImPast] synne.
And quen þis Bretayn *watz bigged* [ImPast] bi þis burn rych,
Bolde *bredden* [Past] þerinne, baret þat *lofden* [Past],
In mony turned tyme tene þat *wroȝten* [Past].
Mo ferlyes on þis folde *han fallen* [ImPres] here oft
þen in any oþer þat I *wot* [Pres], syn þat ilk tyme.
Bot of alle þat here *bult* [Past], of Bretaygne kynges,
Ay *watz* [Past] Arthur þe hendest, as I *haf herde* [ImPres] telle.
Forþi an aunter in erde I *attle* [Pres] to schawe,
þat a selly in siȝt summe men hit *holden* [Pres],
And an outtrage awenture of Arthurez wonderez.
If ȝe *wyl lysten* [FutPres] þis laye bot on littel quile,
I *schal telle* [FutPres] hit as-tit, as I in toun *herde* [Past],
with tonge,
As hit *is stad* [Pres] and stoken
In stori stif and stronge,
With lel letterez loken,
In londe so *hatz ben* [ImPres] longe. (ll. 1–35)[12]

I have indicated a few compound tenses in these lines (imperfect, future) but in dealing with English, one should remember that there are really only two morphologically-marked tenses, past and present. All other temporal distinctions are made lexically and contextually: by adverbials and participles (including using a participle to indicate the aspect of 'pastness' which grammarians call 'imperfect' or 'the continuing past') or by means of conjunction (like *Siþen* in the opening verse or *quen* in the first verse of stanza two). In English still, it is interesting to note that the 'continuing' aspect of a prior event is indicated by either the past or present tense of the verb 'to be' (so, *watz sesed* in line 1 but *is stad* in line 33) or by the present tense of the verb 'to have' (so *han fallen* in line 24). Future time is indicated also by a present tense, this time constructed with an infinite form (*wyl lysten* in line 31). And in lines 8–14, the present tense is used to list a series of discrete

[12] All references are to *Sir Gawain and the Green Knight*, ed. J. R. R. Tolkien and E. V. Gordon, 2nd edn (Oxford, 1968).

events both wholly in the past and non-continuing. These are, of course, fine examples of the 'historical present', used here to relate several unique, completely over-and-done-with, events in history (the founding of Rome, of Lombardy, of Britain, all of which followed upon [*Fro*] the destruction of Troy): in fact, events of exactly the sort for which the 'historical present' is supposed *not* to be used.

In the terms simply of verbal *tense*, therefore, the present is used pragmatically and idiomatically in English to indicate all manner of times, past, present and future. English is by no means alone in this regard: in most languages, present tense verb forms are called upon to indicate all manner of temporal conditions and relationships. Indeed about twenty-five years ago, linguists began challenging the supposition that tenses 'naturally' referred to time at all. A major study in 1975 by C. P. Casparis was called *Tense Without Time*, and the author decried how '[f]or too long the syntagmatic opacity of tense [as to function and meaning] has been concealed behind an inveterate conviction that tense serves in one way or another to express *time*.'[13] He argued that, instead of conceptualizing tense referentially (that is, as signalling extra-linguistic temporal states), one should consider it functionally and pragmatically within actual sentences. Casparis's observation was greatly amplified in a series of studies by Suzanne Fleischman, one of whose principal claims, in her essential 1990 book on *Tense and Narrativity*, 'is that the functions of tense-aspect categories *in narrative* are not limited to the[ir] basic REFERENTIAL meanings . . . the functions of tense-aspect that are exclusive to narrative are specifically pragmatic [textual and expressive] functions.'[14] Casparis and Fleischman, both philologists, focused their study on narrative texts, modern and medieval; in the meantime, beginning in 1983, the philosopher, Paul Ricoeur, had argued in his three-volume study of *Time and Narrative* (to the title of which Fleischman paid tribute in her own) that the only way in which humans can perceive time *is* in narratives, time otherwise is nothing but clock-watching.

The confusion of tense with time of course is ancient, Roman grammarians having used the same word, *tempus*, both for chronological time and for linguistic tense, and our word 'tense' is itself derived, via Old French, from

[13] C. P. Casparis, *Tense Without Time: The Present Tense in Narration* (Bern, 1975), p. 9. Casparis's examples all referred to modern literature, but he insisted upon the functional, non-referential basis of tense use. An early application of this approach, using contemporary oral examples, is D. Schiffrin, 'Tense Variation in Narrative', *Language* 57 (1981), 45–62. See also two important articles by M. Fludernik: 'The Historical Present Tense Yet Again: Tense Switching and Narrative Dynamics in Oral and Quasi-Oral Storytelling', *Text* 11 (1991), 365–98; and 'Narrative Schemata and Temporal Anchoring', *Journal of Literary Semantics* 21 (1992), 118–53. A cogent defence of a pragmatic and historical rather than purely theoretical approach to grammar is P. Hopper, 'Emergent Grammar', *Proceedings of the Berkeley Linguistics Society* 13 (1987), 139–57.

[14] Fleischman, *Tense and Narrativity*, p. 7.

tempus. Augustine plays with this conflation in one of his most-quoted meditations on the nature of time, in the eleventh book of his *Confessiones*:

> Who is he who will tell me how there are not three times [*tempora*, meaning 'tenses'], as we learned when we were boys and as we taught other boys, the past, present, and future, but the present only, because the other two do not exist?[15]

The wry commentary in this question, put by the philosopher Augustine to his former grammarian self, depends upon the semantic conflation of tense with time and indeed it is exactly this analytical incoherence that enables Augustine to realize that, humanly speaking, the extension [*distentio*] of time into past, present, and future is a function of the mind itself.

Instead of conceptualizing tense as referential to an extra-linguistic phenomenon we call time, I would urge with Fleischman that we consider tense pragmatically and functionally, as an element of syntax and morphology alone. '[T]ense', she emphasizes, 'is a formal category of the grammar [of a language]', and it is therefore signalled morphologically,[16] though less so in English than in some other languages, such as Latin. But the temporal localizing of events in a narrative is handled also, and often far more competently, by other means – lexical, contextual and presuppositional. Linguists speak of tense as being inherently 'marked' for time or not. English has only one truly marked tense, the past: 'she left' can only mean that she departed at some time before 'now'.[17] Some languages, though not English, also have tenses specifically marked for the future, and some do not use tense at all to mark time relationships.

The English present tense is unmarked temporally. It can be used to express a speaker's actual present ('I see you over there'); the future ('I'm off' means the same as 'I'll be off'); the past ('You want to know what happened? I was sitting in my office when this guy comes in . . .'); routines ('We meet each Wednesday'); generic statements ('My dog is friendly'); facts ('The sun is a star'); maxims ('All cats are grey at night'). So common is this unmarked characteristic that it seems best analytically to regard the present tense as the basic tense, the one, in the words of the American linguist, Dwight Bolinger, 'to which all other tenses are oriented but which is itself oriented to nothing . . . [It] is "timeless" not in the sense of "eternal" but of "non-committed about time".'[18] Many linguists have agreed that the unmarkedness of the

[15] Augustine, *Confessiones* XI.17: 'quisnam est, qui dicat mihi non esse tria tempora, sicut pueri didicimus puerosque docuimus, praeteritum, praesens et futurum, sed tantum praesens, quoniam illa duo non sunt?' Translations from the Latin are mine, except when indicated otherwise.

[16] Fleischman, *Tense and Narrativity*, p. 18.

[17] The concept of 'markedness' goes back to R. Jacobson and the Prague School: see Fleischman, *Tense and Narrativity*, pp. 52–63.

[18] D. Bolinger, 'More on the Present Tense in English', *Language* 23 (1947), 436. Fludernick,

present is common in ordinary speech; debate continues, however, as to whether (or to what extent) unmarkedness of the present also characterizes *crafted* language, such as epic and romance narrative. I will try to demonstrate now that 'unmarkedness' with respect to time is also true of the present tense in narrative, a matter we can see most readily in the ubiquitous tense switching of medieval texts.

With this in mind, I would like to return briefly to the opening stanzas of *Sir Gawain*. The events narrated by present tense verbs of lines 8–13 (*ricchis, biges, bigynnes, lyftes, settez*) are set in some time past, but one quite indistinct, a time between the destruction of Troy and 'now', the present moment of the narrator in line 26 ('an aunter in erde I *attle* to schawe'). Indeed, the time relations of these two stanzas are entirely carried by causal conjunctions, chiefly 'since' and 'after'; so there are *Siþen* (line 1), *siþen* (line 6), *Fro* (line 8), *synne* (line 19), *quen* (line 19), *syn* (line 23). When things don't occur 'since' and 'after', they occur *Bi syþes* ('By turns', line 16) or *oft* (lines 17 and 22). This temporal mushiness can be unsettling to modern readers, used to carefully dated histories (including fictional histories in novels) and the convention of linear plotting, but it is a cardinal characteristic of medieval narratives. In fact, there is only one specified moment in time in these thirty-six lines of history, and that is the present tense *attle* of line 26 (the present *I wot* of line 23 is generic): 'I intend to show' on this present occasion.

If we consider the present tense to be essentially unmarked in the linguists' sense, the basic tense to which all pasts and futures refer, then I think we have solved the referential problem posed by the 'historical present'. An unmarked tense-without-time can readily be pressed into the service of all sorts of needs, pragmatically, on a moment-by-moment basis. Thus we should not assume with Barthes that the past tense is the cornerstone of narrative and that the narrative present is the problem in need of explaining: rather it seems the case that narrative is basically construed *in the present*. So a very curious cognitive situation pertains to composing narratives. While I would agree that the past is the ordinary *time* of narrative, it seems that the present is the basic *tense* of narrative. Recent linguistic investigations seem indeed to bear out the observation made by St Augustine, that the present is the tense of thought, and of all human cognitive activity, not only of narrative.

A few more instances of tense switching in mid-narrative help to illustrate this point. Here is the Green Knight at Arthur's feast, readying himself for Gawain's axe blow:

'Historical Present', goes a stage further than previous pragmatists in insisting that 'present tense has no meaning, and it is not even conceived of as "timeless", but simply marks a differential value in relation to the past tense' (p. 392). In tense switching, it is the *switch* that is key, she comments; the resulting dynamic of tense-aspect yields patterns of interplay, not static schemes of levels of markedness.

> The grene kny3t vpon grounde grayþely hym *dresses* [Pres],
> A littel lut with þe hede, þe lere he *discouerez* [Pres],
> His longe louelych lokkez he *layd* [Past] ouer his croun,
> *Let* [?Past, ?Pres] the naked nec to þe note schewe.
> Gauan *gripped* [Past] to his ax, and *gederes* [Pres] hit on hy3t,
> þe kay fot on þe folde he before *sette* [Past],
> *Let* [?Past, ?Pres] hit doun ly3tly ly3t on þe naked,
> þat þe scharp of þe schalk *schyndered* [Past] þe bones. (ll. 417–24)

Notice especially the frequent switching between present and past in these lines, all of which describe the same event, even the exact same action (such as Gawain gripping and gathering up his axe). The ambiguous morphology of *Let* (lines 420 and 423) neatly captures the indistinction of tenses in this passage. Yet as readers we have no difficulty understanding that all the action is 'past' *relative to us* and most readers now do not even notice that switching has occurred, nor, I suspect, would the poet's contemporary audience have done so.

In his article on 'Chaucer's Historical Present', to which I alluded earlier, Benson identified a number of uses of the present in narratives which he thought could be classified in terms of various referential (to represent continuous action), expressive (for rhetorical *enargeia*, that is, to 'bring-before-the-eyes' of the reader) or textual functions (for example, to summarize). But of the 1,345 instances identified by Benson in Chaucer of the historical present, 'well over 100 . . . admit of no explanation'.[19] They include single lines such as the following:

> Shrigh*te* [Past] Emelye, and howl*eth* [Pres] Palamoun
> ('The Knight's Tale', l. 2817)

> And forth he go*oth* [Pres], no lenger wol*de* [Past] he tarie
> ('The Pardoner's Tale', l. 851)

> They sw*oren* [Past] and ass*enten* [Pres], every man
> ('The Man of Law's Tale', l. 344)

These lines, in which the two verbs in present and past are repetitive variations of one another, are particularly disappointing candidates for 'post-design justification'. Typical of an older generation of philologists, Fernand Mossé threw up his hands at such lines, classifying them as examples of 'lack of sequence of tense'.[20] More classically-inclined philologists simply spoke of 'ignorance' and 'confusion' on the part of authors and audiences who would permit such a barbarism.

More recently, historians of Old French literature in particular, following

[19] Benson, 'Historical Present', p. 72.
[20] F. Mossé, *Handbook of Middle English* (Baltimore, 1952), p. 98.

Paul Zumthor and Bernard Cerquiglini, have come to view the language of the medieval written texts we now possess as in essence a spoken language, the instrument of a basically oral culture adapted to writing.[21] Fleischman argued, I think compellingly, that the 'disturbing freedom of tenses' in medieval vernacular narratives is an expression of this phenomenon at the level of tense structure. She writes: 'The ontogenesis of the tense-switching phenomenon is, I submit, linked to a particular style of narration . . . the "performed story".'[22] But ultimately I find this explanation inadequate as well. I do not believe that tense switching is a matter always of style, or more particularly, that it 'imitates' the situation of performance. I would reject using any of the criteria of imitation, *mimesis*, to explain the phenomenon of tense switching. Instead, I want to consider it in terms of human cognitive procedure, in particular the creative activity of inventive meditation that the medieval monks called *memoria*.

The most common tense to express past time in later medieval narratives (after the twelfth century or so) is the imperfect. I have mentioned this before: I'd like to take a closer look at the phenomenon now. In Old French literature, the *passé composé* is a tense found infrequently before the twelfth-century narratives of Chrétien de Troyes. A great deal of the *Chanson de Roland* is written in the present tense: indeed it has become a scholarly truism about the *Roland* that its basic narrative tense is the present, its uses of the past tenses are what need explaining. The simple past is used sparingly and under quite restricted circumstances. Instead one commonly gets a sort of compound imperfective tense to express some actions in the past, that is made up of a verb *in the present* plus a past participle: in short, the ancestor of the *passé composé* ('Roland *ad mis* l'olifan a sa buche'). One can say, therefore, that the *passé composé* developed out of the present tense; it is an extension *of the present* into the (indefinite) past.[23] This developmental history seems to me a significant clue about the nature of how we as thinking beings go about conceiving the past (and the future too).

The imperfect aspect of the past is distinctly oriented to the present tense in English as well. We see this in the examples from *Sir Gawain and the Green Knight* already discussed. In the first two stanzas (lines 1–36), I identified two kinds of imperfects, one constructed with the present verb, which I termed

[21] See especially P. Zumthor, *La lettre et la voix: de la 'littérature' medievale* (Paris, 1987); and B. Cerquiglini, *Eloge de la variante: histoire critique de la philologie* (Paris, 1989).

[22] Fleischman, *Tense and Narrativity*, pp. 67–8.

[23] L. Foulet, *Petite syntaxe de l'ancien Français* (Paris, 1919). See also Fleischman, *Tense and Narrativity*, pp. 19–32 and 42–51; and J. Beck, 'On Functional Multiplicity of Tense-Aspect Forms in Old French Narrative', *Romance Philology* 42 (1988), 129–43. On a somewhat related matter, the loss in French of the progressive tense (auxiliary verb plus present participle) and its retention and grammaticalization in English (both, interestingly, phenomena of the eighteenth century), see R. Posner, 'La morphologie "progressive" en Français: aperçu historique', *Linguistique* 34 (1998), 103–10.

'the imperfect present' ImPresent, and one with the Past, the ImPast. In the second stanza are two examples of ImPresent ('as I *haf herde* telle', line 26, constructed with 'have') meaning that the poet knows now the story he heard at some point in the past, the action begun in the present extends, via the past participle, into some time-before-now. 'As hit *is stad* and stoken' (line 33), constructed with the present of 'to be', has the same present orientation: the story is now set down but the down-setting occurred at some time that is past relevant to the present. In the following bit of action from the fox hunt / third bedroom scene, the tense switches are noteworthy, and typical of the narrative as a whole.

> Ofte he [the fox] *watz runnen at* [ImPast], when he out rayked,
> And ofte reled in aȝayn, so Reniarde *watz* [Past] wylé.
> And ȝe he *lad* [Past] hem bi lagmon, þe lorde and his meyny,
> On þis maner bi þe mountes quyle mid-ouer-vnder,
> Whyle þe hende knyȝt at home holsumly *slepes* [Pres]
> Withinne þe comly cortynes, on þe colde morne. (ll. 1727–32)

The past imperfect, constructed with the past of 'be', does not in these lines indicate simply a continuing past but a *routine* past that extends out from and surrounds the present of the narrative, which is the bedroom dialogue between Gawain and the Lady. Renard remains both 'run at' and 'wily' the whole while that Gawain at home wholesomely sleeps, and through his subsequent conversation with the Lady. One could readily analyse *runnen* (line 1727) as simply a predicate adjective parallel with *wylé* (line 1728). By contrast, the simple past *lad* (line 1729) is marked relative to a specific time reference to the hour of *mid-over-undren*, 'somewhere around noon' the present tense when Gawain rouses from his sleep. In other words, it is the present to which the various past tenses are related. Again, the present is the ordinary tense of thinking, the tense of composing and inventing. And a morphological switch to the past tense signals a relationship to this present that must acquire its meaning contextually and presuppositionally rather than formally. Modern literary convention, not nature, decreed that stories ought to be narrated in the past, just as ancient grammarians decreed that tenses must be temporally sequenced, thereby prohibiting tense switching in Latin composition. Only convention makes it meaningful, and the conventions of narrative we now know have taken hold since medieval literature was composed.[24]

As a morphological phenomenon, at least in English, tense switching is

[24] Narratology as a field is for the most part predicated on post-medieval literature. Even the so-called 'natural' story-telling of contemporary taped oral narratives uses the conventions of its genre. One of the most evident of these now is using past tense as the unmarked, backgrounded, and thus 'normal' tense of narrative. But this is a modern convention, whose ascendency has been observed to coincide with the dominant popularity of the novel.

unremarkable both in spoken and written language and we often overlook it. To return for one moment to the example I gave earlier of the present used for events in past time. 'You want to know what happened? I'm sitting in my office when this guy comes in . . .' or 'I was sitting in my office when this guy comes in . . .' or 'I was sitting in my office when this guy came in . . .' or 'I'm sitting in my office when this guy came in . . .'. All four variants mean the same thing: tellingly, unless they are particularly made aware of the tense variations, most people simply don't either hear them or see them. I would thus analyse Chaucer's arbitrary tense switching into the present ('Shrighte Emelye, and howleth Palamoun'; 'They sworen and assenten, every man') as simply non-meaningful, a twitch that both the poet and his audiences read right over, untroubled and unnoticing. But that we can pass over them so readily is, I think, a sign that tense use is not always referential to time.

To summarize thus far, tense conceptualizes 'the location of situations predicated in a sentence or discourse relative to a reference time . . . In contrast to other grammatical categories associated with the verb (aspect, voice, mood) . . . tense is *relational*.'[25] In order to conceive of *time*, in other words, we use a notion of *space*, for 'location' is a spatial quality. This observation is central in Augustine's examination of time too, and indeed I wonder if his experience as a teacher of rhetoric wasn't more important to his acknowledged philosophic breakthrough with this problem than his reading of those famous 'Platonic books'. When Augustine first raises the problem of time in book XI.18 of the *Confessiones*, he asks himself:

> [I]f there be times past, and times to come; fain would I know *where* they be: which yet if I be not able to conceive, yet thus much I know, that *wheresoever* they now be, they are not *there* future or past, but present.[26]

He asks not *what* past and future might be, but *where* they might be (*ubi sint?*) His ironic puzzlement in the subsequent chapters over how time can be measured is an aspect of this same insight.

Only things having dimension can be measured; the present is simply a point of experience, without extension. Yet clearly human beings can measure time; they have a good sense of relative duration. Augustine's example is the first verse of the Ambrosian hymn, *Deus creator omnium*, which contains, alternately, four long and four short syllables. Ricoeur discussed this example at length in his *Time and Narrative*: I have no desire to repeat his analysis now.[27] Basically, Augustine observes that he can measure duration only relatively (short and long) and only through the

[25] Fleischman, *Tense and Narrativity*, p. 15.

[26] 'si enim sunt futura et praeterita, volo scire, ubi sint. quod si nondum valeo, scio tamen, ubicumque sunt, non ibi ea futura esse aut praeterita, sed praesentia.' Augustine, *Confessiones* XI.18.

[27] Ricoeur, *Time and Narrative*, I, 13–22.

simultaneous activities of his mind, the recollection of a memory, the perception of the present sound, and the 'expectation' of the syllables yet unsaid. Therefore, one cannot truly speak of past and future as 'things', but only as mentally-constructed relationships to the present:

> Nor do we properly say, there be three times, past, present, and to come; but perchance it might be properly said, there be three times: a present time of past things; a present time of present things; and a present time of future things. For indeed three such as these in our souls there be; and otherwise I do not see them. The present time of past things is our memory [*memoria*]; the present time of present things is our sight [*contuitus*]; the present time of future things our expectation.[28]

He concludes, in chapter 27: 'in you, my soul, I measure *my* times/tenses [*in te, anime meus, tempora mea metior*]'. Notice the use of the personal possessive pronoun, for this is what will bring me to the next logically incoherent but commonly understood concept I want to explore, that of 'collective memory'. The confusion inherent in this concept is, I will argue, bound up in the paradox that in order to perceive past and future, one must plot time spatially in the present. And in a specific kind of present at that, the present of the mind recollecting, that of inventive *memoria*.

Augustine appears to distinguish three mental activities as he discusses how past and future are related to the present of thinking. But his example of the hymn verse is one that privileges recollection: I measure duration in the present by *recollecting* a syllable in my memory and comparing it with my present perception; my present perception immediately passes into my memory; I can expect the words to come only because I have already memorized them, otherwise I could not 'expect' them. So the whole problem of *tempora*, 'times', becomes a problem of memory, of *where* things are in the mind, a problem of structural relationships.

Human time, for Augustine, appears to behave rather like the present tense, in that past and future exist syntactically, as a relationship to an otherwise temporally unmarked 'present'. The 'past of the present', 'the future of the present', 'the present of the present', these striking phrases all relate temporality to the present alone. Indeed English grammar has exactly the tenses (*tempora*) Augustine plays with here: a past of the present (*I haf herde telle*) and a future of the present (*I schal telle hit*), both of which *locate* events in relation to the present.

Commenting on the rhetorical memory rules in the ancient treatise, *Rhetorica ad Herennium*, Albertus Magnus raises as an objection, that the Roman treatise gives only spatial and locational rules for recollection, and

[28] 'fortasse proprie diceretur: tempora sunt tria, praesens de praeteritis, praesens de praesentibus, praesens de futuris. sunt enim haec in anima tria quaedam, et alibi non video: praesens de praeteritis memoria, praesens de praesentibus contuitis, praesens de futuris expectatio.' Augustine, *Confessiones* XI.20.

never suggests that memories are time-based, 'of things that are in the past', as Aristotle had said. Albertus replies to this objection by observing that, since all memories are of the past their 'pastness' does not serve to distinguish among them, and therefore that they need to be 'marked' in other ways: by characteristics such as place, size, movement, relation to one another. One cannot know how long in the past a memory is, it must be distinguished and 'located' by other means. Indeed Albertus goes on to observe that memory can be strengthened and trained only because it is spatial and locational, the art of memory is an art of *places*.[29] Its usefulness to an orator composing, whether on the spot or in advance, is particularly as an art of common places, what the ancient textbooks called *memoria rerum*.

Such mental emplacement of things remembered at some time in the past allows the orator to recall remembered things, of all sorts, in the present. So, while the time of memories may be the past, of necessity, as Aristotle observes, because what our brains store is an image or trace, and some time must have passed between the perception and the formation and storage of its 'vestige', the tense of remembering is present. It is this recollective present, the re-presentation of memories, that Augustine meditates on in Book XI of the *Confessiones*.

In Aristotle's account 'Concerning Memory and Recollection', he distinguishes quite rigorously between the two. Memory, he says, is an *affectus*, an experience recorded in the soul; it is composed of an image of some sort and our response to it, it 'stamps' us physiologically. Recollection uses memory images, but it is a species of thinking; he even says at one point that recollection is a process of inference, like reasoning.[30] But although Aristotle starts out distinguishing the two carefully, using *mneme* and related words for 'memory' and *anamnesis* and related words for recollection, he also uses *mneme* when he is plainly talking about recollection. This strongly suggests that he fell easily into a common Greek usage (which we also have) of regarding recollection as a form of memory and in so doing, of course, we bring a past into a present activity (remembering).

In our everyday experience now we remember two sorts of memories: we remember our own experiences and we remember things that happened in 'the past'. Julia Annas, a historian of Greek philosophy, distinguishes these as remembering 'the past' and remembering 'my past'.[31] We distinguish them

[29] Albertus analyses this in his discussion of the nature of prudence in *De bono* 4, qu. 2, ad 2, especially objection and response 7, in *Opera omnia*, ed. H. Kühle, C. Feckes *et al.* (Aschendorff, 1951). A translation appears in M. Carruthers, *The Book of Memory: a Study of Memory in Medieval Culture* (Cambridge, 1990), pp. 270–1 and 277.

[30] Aristotle's treatise 'On Memory and Recollection' has been translated with extensive commentary and notes by R. Sorabji, *Aristotle on Memory* (Providence, RI, 1972).

[31] J. Annas, 'Aristotle on Memory and the Self', in *Essays on Aristotle's* De Anima, ed. M. Nussbaum and A. O. Rorty (Oxford, 1992), pp. 297–311; and J. Annas, 'Memories are Made of This and That', *London Review of Books* 14:9 (1992), 21–2.

all the time. Consider these two very common sentence patterns: 'I remember Paris' and 'I remember that Paris became the capital of France.' Facts, such as that Paris became the capital of France, may be cast in a past tense to indicate that the event happened in the past but in effect such historical facts are without time. That Paris became the capital at a certain point is as true now as it was then, and will always be true. And it is true for everyone: it is a fact (note the present tense).

If I say now 'I remember that Chaucer wrote *Troilus and Criseyde*' you would understand me to be saying that I recall that Chaucer wrote *Troilus and Criseyde* because I learned this fact at some time in my past. If I were to say to you, however, 'I remember Chaucer writing *Troilus and Criseyde*', you would quite rightly look at me strangely. We make firm boundaries between what Annas calls 'non-personal' (or factual) and 'personal' memory. But sentences of the type 'I remember Chaucer writing *Troilus*' are really quite common in medieval narratives. Take, for example, the passage from *Sir Gawain and the Green Knight* wherein the poet tells of when the Green Knight came into Arthur's court: 'Half etayn in erde I *hope* þat he were, / But mon most I algate *mynn* hym to bene.' (ll. 140–1). The poet remembers, *I mynn*, him to be at least the biggest of men, if not half a giant. This use of *mynn* is usually glossed as 'judge' rather than 'remember' but that is an effort to square the remark with modern ideas. Combined with *I hope*, in the previous line, this comment puts the poet in the position of a present-tense observer, now recalling a personal memory image of what he once experienced. Moreover, there is also a succinct extension in these two lines from the present *tense* of both verbs into the implied future *time* of expectation (*I hope*) and past time of memory (*I mynn*).

Chaucer plays off these two kinds of remembering, non-personal (that of history) and personal, to wonderful effect throughout his *Troilus*, using the person of his narrator. Modern readers have been struck by the narrator's intense engagement with his story, the way that he enters the action as though he were present. And the narrator has often been criticized for doing this, for 'forgetting' (that is the verb critics have used) the story he is supposed to be telling and lacking a historian's proper distance. Here is an example (nearly all of the obvious instances involve Criseyde, as this one does):

> Criseyde was this lady name al right.
> *As to my doom*, in al Troies cite
> Nas non so fair, forpassynge every wight,
> So aungelik was hir natif beaute (i, 99–102)

And another:

> Criseyde mene was of hire stature;
> Therto of shap, of face, and ek of cheere,

> Ther myghte ben no fairer creature.
> And ofte tymes this was hire manere:
> To gon ytressed with hire heres clere
> Doun by hire coler at hire bak byhynde,
> Which with a thred of gold she wolde bynde;
>
> And, save hire browes joyneden yfeere,
> Ther nas no lak, in aught *I kan espien*. (v, 806–14)

Here again the present reaches into the past of the story. The narrator judges now, *As to my doom* (i, 100). And in book v, line 814, we get a fine instance of tense switching from the past of the description to the present, *I kan espien*. It has often been observed that the time of this set of descriptions in book v is separated from the specific events of the history, occurring between two simple pasts, Diomede '[t]houghte in his herte' (v, 796) and '[i]t fel that after, on the tenthe day' (v, 842), in a generalized and seemingly timeless narrative past. But it is notable in this part of the description how tense is further shifted from the generalized past of *was*, to the conditional *myghte ben*, to the routinized *ofte tymes*, and finally to the simple present of *I kan espien*. And the emphatic personal pronoun, I, is the focus of this tense switch. I don't think this is because Chaucer wants to show moral failings in the narrator, however, but because he is here (as so often in this painful poem) stretching himself out to encompass all the times of this story in their relationships: in short, to place himself, and thus to remember.

A sense of time is always personal, 'my times', *tempora mea*, as Augustine says. 'Past' simply as pastness, as Albertus Magnus observed, is a non-significant characteristic because it is unrelated and unspecified, beyond its being 'past'. It is common in mnemotechnic to personalize all the *dicta et facta* one learns in school *memorabilia*, 'for remembering'. To use Annas's terms, in medieval memory-work one consciously embeds non-personal memories within personal ones. History, 'the past', thus is marked in some way also as 'my past'. In some elementary advice written by Hugh of St Victor in the early twelfth century for novices needing to learn the Psalms and also the essential time-line of biblical history, facts which were among the fundamentals of a regular clerical education, Hugh urges the novices to set the facts within their own experience of learning them, as a recollective tool, to store a fact within a personally marked mental locale. Thus, one should pay attention to the voice and gestures of the teacher, the time of day, the weather, and, in the case of written texts which one is memorizing silently, to the appearance of the codex page on which the writing occurs, the colours of the inks, the decorations, whether the page is hair- or skin-side.[32] Of course one should also mark one's memories with some sort of mental notation. And

[32] This advice is in Hugh's preface to a chronicle of biblical history which he made for the novices of St Victor; a translation of it is in Carruthers, *Book*, pp. 261–6.

overall careful and caring attention must be concentrated on the task. To inhere in my memory, and thus to be available for my thought, even (perhaps especially) the facts I learn need to be made into my own memories, so that they can be located again because I have had some*where* to put them in my mind.

It is time now to look at one of the most quoted and, I suspect, least completely understood of all definitions of time. But I'm going to make a stab at it, as a way of trying to bring the various strands of my paper together. Because the translation is itself the problem to analyse, I will give the Latin text: 'inde mihi visum est nihil esse aliud tempus quam distentionem: sed cuius rei nescio, et mirum, si non est ipsius animi.'[33] The spatial nature of human time perception, and time-based understanding, is encapsulated by Augustine in this definition of time as a *distentio*, literally 'a stretching', of the mind. Augustine's noun *distentio* has caused much commentary. The word was used very rarely in Latin at this time; in his translation of the *Confessions*, H. Chadwick has suggested that Augustine coined the word himself, though the *Oxford Latin Dictionary* lists a couple of uses from the first century of the Empire. It is usually translated as 'extension', 'time is an extension of the mind', an expression innocuously tame in modern English. The seventeenth-century translation by William Watts printed in the Loeb edition is 'a stretching out in length', which gives a better sense that the mind is actively moving itself.[34] But why did Augustine not use *extensio* if he simply meant 'stretching out', a commoner and less fraught word, which in fact he uses in contrast to *distentio* when he re-uses these nouns in his extraordinary closing prayer two chapters later.

Distendo, the verb from which this noun is made, means 'swell', 'stress' and 'strain' as well as simply 'extend': the earlier uses of *distentio* are glossed by the *Oxford Latin Dictionary* as 'strain' and 'spasm, distortion'. Evidently, Augustine meant to emphasize that the sort of mental stretching involved in time-making was painful; it hurt. In a note, Chadwick writes that:

> [I]n Augustine this psychological experience of the spreading out of the soul in successiveness and in diverse directions is a painful and anxious experience, so that he can speak of salvation as deliverance from time.[35]

Thus in her new translation, Maria Boulding selects 'tension' for *distentio*, and notes, like Chadwick, that the word 'carries a subjective charge of distraction or anxiety of soul, like being stretched on a rack'.[36] Time-making, like all other

[33] 'wherefore it seems to me that time is nothing else than stretching: but of what material I know not, and I wonder, if it is not of the mind itself.' Augustine, *Confessiones* XI.27.

[34] Augustine, *St Augustine's Confessions*, ed. P. Knöll, trans. W. Watts (London, 1912). As archaic as the Watts translation is, it often has a vigour of phrase, closer to Augustine's Latin, that more recent ones lack.

[35] Augustine, *The Confessions*, trans. H. Chadwick (Oxford, 1991), p. 240, n. 27.

[36] Augustine, *The Confessions*, trans. Boulding, p. 304, n. 82.

kinds of thought-making, involves the 'tension of consciousness itself', in her fine phrase.

Such painful tension is life itself. As Augustine writes in summary: 'ecce distentio est vita mea', 'my life is straining'. At this point, he contrasts *distentio* to *extentio*, meaning also 'stretching out', but stretching and smoothing, extending in a straight line, towards the timeless and eternal. He then prays that:

> I may be gathered up from my old days following the One, having forgotten my pasts: not distorted [*distentus*] towards those things which are future and transitory, but made straight [*extentus*] towards those that are from before [*ante*] all things, not by straining [*non secundum distentionem*] but by focusing [*sed secundum intentionem*] I follow on towards the victory palm of my heavenly calling.[37]

This present tense is not timeless in the sense of being uncommited as to time, straining into the past and the future, but it is timeless in the sense of being eternal; it therefore has no relationship, being simply 'before' (*ante*) all.

It is usual to emphasize the bad qualities of *distentio*, and certainly Augustine intends them. But we must also remember, for he certainly did, that such restless anxiety and strain were considered necessary to creative action, the activity of knowing itself. We now assume that all bad things, especially psychological pain, are always to be avoided, but this assumption was not shared. I have written extensively in *The Craft of Thought* on the highly emotional nature of monastic meditation, each stage of the way marked emotionally as a series of stages beginning in intense anxiety and fear and then proceeding through a variety of other emotions. Augustine himself wrote about these stages of meditation, in *De doctrina christiana*, as self-induced, beginning with a meditation on your own death until in your fear, you have raised such gooseflesh on your skin that, throwing yourself in terror upon the cross, you will stick to it as though your flesh had broken out in little nails.[38]

It is against this background that I understand Chaucer's account of beginning to compose his *Troilus*. I don't think we should see his tenderheartedness as weak-headedness, nor as anything exceptional in a medieval author. He does emphasize the great anxiety with which he sets forth on his great task:

[37] 'ecce distentio est uita mea, . . . et a ueteribus diebus conligar sequens unum, praeterita oblitus, non in ea quae futura et transitura sunt, sed in ea quae ante sunt non distentus, sed extentus, non secundum distentionem, sed secundum intentionem sequor ad palmam supernae uocationis.' Augustine, *Confessiones* XI.29.

[38] Augustine, *De doctrina christiana*, ed. J. Martin, CC Series Latina 32 (Turnhout, 1962), II.vii.9. See the discussion in Carruthers, *The Craft of Thought: Meditation, Rhetoric, and the Making of Images, 400–1200* (Cambridge, 1998), esp. pp. 60–115.

> The double sorwe of Troilus to tellen,
> That was the kyng Priamus sone of Troye,
> In lovynge, how his aventures fellen
> Fro wo to wele, and after out of joie,
> My purpos is, er that I parte fro ye.
> Thesiphone, thow help me for t'endite
> Thise woful vers, that wepen as I write.
>
> To the clepe I, thow goddesse of torment,
> Thow cruwel Furie, sorwynge evere in peyne,
> Help me, that am the sorwful instrument,
> That helpeth loveres, as I kan, to pleyne;
> For wel sit it, the sothe for to seyne,
> A woful wight to han a drery feere,
> And to a sorwful tale, a sory chere. (i, 1–14)

The world of 'the past', of Troy and Troilus's double sorrow, is made into that of 'my past' in these lines, as the common *wel wist* material of *the* past (i, 57), is made into *my* times, the tense wherein the narrator and his verses weep together. Thesiphone, sorrowing ever in pain, the goddess of torment, may seem an odd choice of Muse to us, but she is absolutely appropriate for this story remembered from old books, about the past and my past, the need for the one to be embedded within the experience of the other, and the pain of creation that rightly accompanies a poet's effort to stretch out his soul in order to gather his subject matter into his own times and remember Troilus by remaking him.

8

The Sense of Anachronism from Petrarch to Poussin

Peter Burke

This essay does not attempt to encompass the whole range of images of the past current in Europe from Petrarch to Poussin.[1] It is deliberately limited to a certain view of the past, especially its relation to the present. This view has been described in terms of 'historical perspective', 'antiquarian feeling', 'archaeologizing spirit', 'historicism' (*Historismus*) or *Geschichtsbewusstsein*. It might be equally appropriate to view the tendency from the opposite end and to speak of 'awareness of innovation'.

The variety of names given to this tendency by modern scholars suggests a certain awkwardness in pinning down this elusive butterfly. The problem is a recurrent one in the history of mentalities, concerned as this approach is with the implicit, the unspoken, with what is so obvious (at a given time, in a given place, to a given social group) that it is taken for granted. Perceptions of space are an obvious example. So are perceptions of time, or what the French call *régimes d'historicité*.

THE DISCOVERY OF THE RENAISSANCE SENSE OF THE PAST

The central argument of this essay, originally put forward in 1968, and here at once reiterated, extended and qualified, is that a new sense of the past was felt and expressed by some scholars and artists in the fifteenth and sixteenth centuries, originally and especially in Italy. Difficult to define, the new sense of the past is most clearly illustrated by the fact that some people of the time defined their own age as the Renaissance and contrasted it with what they were the first to call the Middle Ages. Like a new nation, the humanists in particular constructed a collective identity by contrasting themselves with the 'other'. They invented the term 'Middle Ages' (*medium tempus, medium aevum*), just as they invented the terms 'Dark Ages',

[1] This article represents a third version of my opinions on the subject. Earlier attempts to analyse the phenomenon were published as 'The Sense of Historical Perspective in Renaissance Italy', *Journal of World History* 11 (1968), 615–32; and 'The Renaissance Sense of the Past Revisited', *Culture and History* 13 (1994), 42–56.

'Gothic', and 'schoolmen' (*scholastici*), all labels which remained influential for centuries.[2]

In short, the humanists considered the past (ancient and medieval) to be different in quality from the present, rather than more and more of the same thing. In the case of their attitudes to antiquity one might speak of 'nostalgic distance', an awareness of difference joined with admiration and a desire to annihilate that difference. In the case of humanist attitudes to the Middle Ages, on the other hand, it might be better to speak of 'ironic distance'. In practice, however the range of attitudes to the Middle Ages expressed by sixteenth-century Italians runs from the affectionate irony with which Ariosto rewrote the medieval romance in his *Orlando Furioso* to the contempt with which Vasari (at times, at least) dismissed Byzantine and Gothic art (which he described as the *maniera greca* and *maniera tedesca* respectively).

Like other forms of consciousness, historical consciousness fascinated Hegel, who discussed it in his *Philosophy of History*.[3] There were doubtless good reasons for the rise of this consciousness of consciousness in the nineteenth century, especially in the German-speaking world. Burckhardt did not seriously discuss this aspect of the Renaissance, perhaps because of his well-known distaste for Hegel's philosophy of history.[4] Later German scholars, however, amply compensated for this omission. Karl Lamprecht, for example, emphasized the lack of a sense of historical distance in the Middle Ages and its discovery in the age of Columbus, a simultaneous 'Ausdehnung des raümlichen Horizonts' and 'Ausdehnung des zeitlichen Horizonts'.[5] Erich Auerbach also contrasted the earlier medieval references to 'chivalry' or 'vassals' in biblical times with the rise of a sense of distance in the age of humanism.[6] Literary scholars in France and England made similar points about anachronisms in medieval literature, or for that matter in Shakespeare. Gaston Paris, for example, argued that medieval people thought that the world had always been the way they saw it.[7]

However, the idea of anachronism was studied with most precision by German historians of art, notably by Aby Warburg, Fritz Saxl, and Erwin Panofsky. Warburg noted Botticelli's concern with period costume: 'eine

[2] L. Varga, *Das Schlagwort vom Finsteren Mittelalter* (Vienna, 1932); and P. Frankl, *The Gothic: Literary Sources and Interpretations through Eight Centuries* (Princeton, 1960).

[3] G. W. F. Hegel, *Vorlesungen über die Philosophie der Geschichte* (Stuttgart, 1961), pp. 549–52.

[4] J. Burckhardt, *The Civilization of the Renaissance in Italy*, trans. S. G. C. Middlemore (Harmondsworth, 1990).

[5] K. Lamprecht, *Deutsche Geschichte*, 12 vols. (Freiburg, 1904–10), II, part 2, 3–56.

[6] E. Auerbach, *Mimesis: the Representation of Reality in Literature*, trans. W. R. Trask (New York, 1950), p. 140.

[7] P. Stapfer, *Shakespeare et l'antiquité* (Paris, 1879), pp. 93–109; M. W. MacCallum, *Shakespeare's Roman Plays* (London, 1910); and G. Paris, *La littérature français au Moyen Age* (Paris, 1888), p. 30.

gewisse archäologische Treue des Kostüms'.[8] Saxl focused on the historical interests of Mantegna and his friendship with the antiquarian Fra Feliciano of Verona.[9] Perhaps the most famous discussion of the topic, however, is Panofsky's comparison between the sense of historical perspective and the pictorial perspective of the Renaissance.[10]

It was perhaps no accident that art historians led the way in this investigation, for art is at once an essential and a treacherous source for historians of mentalities. It is treacherous (from the point of view of a historian interested in evidence) because it has its own conventions, because it follows a curve of internal development as well as reacting to the cultural environment and finally because it is necessary to distinguish occasions when artists are expressing their individual attitudes from others when they are simply articulating the attitudes of their patrons.

On the other hand, art is an essential source for historians of mentalities, concerned as they are with unspoken assumptions as much as with conscious attitudes, above all because an image is necessarily explicit on issues which may be evaded more easily in texts. The history of a concern with the history of costume may seem a trivial subject, but awareness of changes in costume is the outward sign of the sense of historical distance which is under investigation here. As Freud, Morelli and Holmes (not to mention Carlo Ginzburg) have taught us, apparently trivial phenomena may offer valuable clues to what is deeper. Or, as Warburg used to say, 'Der liebe Gott steckt im Detail.'[11]

A little later, some intellectual historians took up the theme of the sense of the past, particularly among renaissance humanists. Hans Baron's article on the 'awakening' of historical consciousness in fifteenth-century Italy, and Theodor Mommsen's study of Petrarch's view of history were among the earliest explorations of this topic.[12] Particular attention was given to the way in which the humanists, for whom philology was such a central concern, became increasingly aware of changes in language over time. The most famous example is that of Lorenzo Valla in his exposure of the so-called *Donation of Constantine* as a forgery. What was new in Valla's demonstration was not his scepticism about the *Donation*, which had been shared by earlier writers, such as William of Ockham, but the weight which he gave in this

[8] A. M. Warburg, *Sandro Botticellis 'Geburt der Venus' und 'Frühling'*, 2 vols. (Leipzig, 1932), I, 3–58.

[9] F. Saxl, *A Heritage of Images*, ed. H. Honour and J. Fleming (Harmondsworth, 1970), pp. 57–70.

[10] E. Panofsky, 'The First Page of Vasaris *Libro*', in *Meaning in the Visual Arts: Papers in and on Art History* (New York, 1957), pp. 169–225.

[11] C. Ginzburg, 'Clues', in *Myths, Emblems, Clues*, trans. J. Tedeschi and A. C. Tedeschi (London, 1990), pp. 96–125.

[12] H. Baron, 'Das Erwachen des Historischen Denkens', *Historische Zeitschrift* 147 (1932–3), 5–20; and T. E. Mommsen, 'Petrarch's Conception of the Dark Ages', in *Medieval and Renaissance Studies*, ed. E. F. Rice, Jr (Ithaca, 1959), pp. 106–29.

demonstration to the analysis of its language, a form of Latin which he showed to be centuries later than Constantine.[13] In short, Valla had an acute sense of what we call 'anachronism'.

CRITICISMS AND PROBLEMS

The thesis of the rise of 'the renaissance sense of the past', which once seemed virtually obvious, may be and has been criticized from two opposite (indeed, apparently inconsistent) points of view. On one side some medievalists claim that 'their' period too was characterized by a sense of the past, which cannot be described simply in terms of absence. They note that some fourteenth- or fifteenth-century examples might just as well be credited to the Middle Ages as to the Renaissance. One scholar goes so far as to write of 'the deep sense of history' of medieval writers who discuss the *Donation of Constantine*.[14]

On the other side, we find the argument, or in some cases the assumption that the sense of historical distance is later than the Renaissance. The humanist belief that it was possible to revive antiquity, like the artistic convention of representing fifteenth- or sixteenth-century individuals in ancient Roman costume, contradicts the thesis of the renaissance sense of the past, so does the humanist interest in exemplarity.[15] There was no idea of the development of Christian doctrine: doctrine was supposed to be unchanging.[16] Again, in the context of renaissance political thought, a recent study emphasized 'the extent to which' the humanists 'lack any sense of the past as a foreign country'.[17] If these arguments are valid, we will have to look for this sense of distance in a later period, such as the late eighteenth century, as Meinecke did.[18]

It will be obvious enough that to decide these issues it is not sufficient to carry out more research. We also need to pose more precise questions and to draw more distinctions, three in particular.

[13] F. Gaeta, *Lorenzo Valla: filologia e storia nell'umanesimo italiano* (Naples, 1955); D. Maffei, *Gli inizi dell'umanesimo giuridico* (Milan, 1956); and D. R. Kelley, *Foundations of Modern Historical Scholarship* (New York, 1970), ch. 1.

[14] R. W. Southern, 'The Sense of the Past', *Transactions of the Royal Historical Society* 5th s. 23 (1973), 243–63; B. Guénée, 'Y a-t-il une historiographie médiévale?', *Revue Historique* 524 (1977), 261–75 (pp. 261–3); and R. Black, 'The Donation of Constantine: A New Source for the Concept of the Renaissance', in *Language and Images of Renaissance Italy*, ed. A. Brown (Oxford, 1995), pp. 51–86 (esp. pp. 64 and 84).

[15] T. Hampton, *Writing from History: the Rhetoric of Exemplarity in Renaissance Literature* (Ithaca, 1990).

[16] W. O. Chadwick, *From Bossuet to Newman: the Idea of Doctrinal Development*, 2nd edn (Cambridge, 1987), ch. 1.

[17] Q. Skinner, *Reason and Rhetoric in the Philosophy of Hobbes* (Cambridge, 1996), p. 40.

[18] F. Meinecke, *Die Entstehung des Historismus*, 2nd edn (Munich, 1959).

1. To begin with asking a sociological question: *whose* sense of the past? What confuses the issue is the fact that the term 'Renaissance' is traditionally used to refer to both a movement and a period. It is unlikely that anyone would argue that the majority of the population of renaissance Italy shared in the new sense of the past. It is probable that as late as the nineteenth century, European peasants lacked a sense of historical distance (a conclusion which may be illustrated, if not proved, by the peasant paintings from Dalarna in Sweden representing biblical characters in modern dress). The new sense of the past was characteristic of a group of humanists and artists, a small group in the fifteenth century but a substantial one in the sixteenth.

2. Even within the group of Italian artists and humanists, the sense of anachronism was not always present. Aby Warburg drew attention to the example of the artist Pesellino, who according to him lacked any sense of distance between past and present ('keine Distanz zwischen heute und Vergangenheit gelten lässt'), and to that of a humanist who made no objection to the representation of Caesar on a tapestry in the form of a knight. More recently, a study of the renaissance churchman Giles of Viterbo emphasized his view of the past as 'homogeneously continuous with the present, culturally undifferentiated from it'.[19] It would appear that anachronism and a sense of anachronism could coexist within the same group or even within the same individual.

In order to understand this phenomenon it might be prudent to refine the crude distinction between the presence and the absence of historical consciousness, and to try instead to identify degrees of historical distance, or a more or less acute awareness of anachronism. Modern scholars have a more acute sense of historical distance than renaissance humanists, and as a result the humanists may appear to them to lack any sense of distance. However, the humanists had more sense of distance than their medieval predecessors.

3. In any case, there was a tension between two characteristically renaissance attitudes, not only the sense of distance between the present and the admired classical past, but also the confidence in the possibility of annihilating that sense of distance and reviving the dead. As a recent commentator has put it:

> One set of humanists seeks to make the ancient world live again, assuming its undimmed relevance and unproblematic accessibility; another set seeks to put the ancient texts back into their own time, admitting that reconstruction of the past is difficult and that success may reveal the irrelevance of ancient experience and precept to modern problems.[20]

[19] J. W. O'Malley, *Giles of Viterbo on Church and Reform* (Leiden, 1968), p. 182.
[20] A. Grafton, *Defenders of the Text* (Princeton, 1991), pp. 26–7.

The tension between the two attitudes is revealed most explicitly in the art of the period. Renaissance artists were more sharply aware than medieval artists had been that the Romans wore different clothes from their successors. Some of them gave up representing Romans in modern dress. On the other hand, they had no apparent inhibitions about representing moderns in Roman dress, making them dignified by giving them the appearance of the admired and exemplary ancients, a practice still common in the late eighteenth century, as the controversy about Benjamin West's *Death of General Wolfe* reminds us.[21]

In other words, the label 'the renaissance sense of the past', like so many historical labels, is not altogether satisfactory. It is useful in reminding us of the centrality of a new sense of the past in that movement, but misleading in its implication that a single sense of the past was characteristic of the whole period.

There is yet another problem which must be confronted here. How is the historian to characterize medieval attitudes to the past? The traditional response within Renaissance Studies, as we have seen, is to describe them negatively, in terms of a lack of awareness. The problem is not of course specific to our topic; this is what Burckhardt did when he wrote about individualism. The French historian Philippe Ariès described the Middle Ages in a similarly negative fashion in his famous study of the sense of childhood ('le sentiment de l'enfance').[22] In each of these cases, powerful objections have been raised against this characterization. I still believe that important, if gradual, changes took place in attitudes to the past in the fifteenth century in particular. However, I would now prefer to follow Sir Richard Southern and to speak of a medieval 'sense of the past' being replaced by another, later one.[23]

A last problem or cluster of problems demand to be mentioned here. Supposing that attitudes to the past changed, why did these changes take place? There are at least two kinds of explanation to consider. The first possibility is that a sense of discontinuity between past and present is a reaction to a major social or political change such as the Reformation or the French Revolution. The second is an explanation in terms of the media of communication, along the lines of the argument advanced by the anthropologist Jack Goody in the 1960s to the effect that the use of writing, which fixed certain views of the past and inhibited what he calls 'structural amnesia', encouraged an 'awareness of the past as different from the present'.

[21] E. Wind, 'The Revolution of History Painting', *Journal of the Warburg and Courtauld Institutes* 2 (1938–9), 116–27.

[22] P. Ariès, *L'enfant et la vie familiale sous l'ancien régime* (Paris, 1960).

[23] Southern, 'Sense of the Past'.

The shift from 'memory' to 'written record' in medieval England exemplifies Goody's point very clearly.[24]

THE RISE OF A SENSE OF THE PAST

With the distinctions drawn above in mind, the rest of this essay will discuss examples of the sense of the past in more or less chronological order. The only problem is when to begin. A generation ago, a study of the Anglo-Saxon sense of the past emphasized the lack of any sense of anachronism at that time.[25] A recent study of the ninth-century monk-historian Notker rejects the idea that he had a pre-literate mentality, while admitting that 'to a large extent', Notker's chronicle 'lacks any sense of chronological development or historical time'.[26]

For the moment at least, the year 1000 or thereabouts may offer a convenient starting-point. Patrick Geary has recently suggested that:

> Across the Carolingian world, people of the eleventh century seemed as interested in questions of continuity and discontinuity as subsequent historians, with some emphasising the novelty of their age and others its continuity with the past.[27]

It is the latter view which is supported by the evidence of images, which represent figures from the past such as the Roman soldiers arresting Christ or sleeping at the Sepulchre during the Resurrection in the clothes of the artist's own time. For example, an illustration in a manuscript of about 1025 from Gembloux or Liège shows the soldiers in eleventh-century coats of mail, reminiscent of the warriors on the Bayeux Tapestry.[28] This convention does not change in the twelfth century.

Literature tells a similar story, at least on the surface. In his romances Chrétien of Troyes, for example, described the Greeks and Romans as fighting in tournaments like medieval knights. Today, there is a tendency to describe Chrétien and his peers not as making a naïf assumption of continuity but as consciously mediating between past and present, as 'suppressing' distance by 'an act of intellectual will'.[29] In a similar fashion,

[24] J. Goody, *Literacy in Traditional Societies* (Cambridge, 1968), pp. 34, 49 and 56; and M. T. Clanchy, *From Memory to Written Record: England 1066–1307* (London, 1979).

[25] M. Hunter, 'The Sense of the Past in Anglo-Saxon England', *Anglo-Saxon England* 3 (1973), 29–50.

[26] M. Innes, 'Memory, Orality and Literacy in an Early Medieval Society', *Past and Present* 158 (1998), 3–36. A further study by Dr Innes of the early medieval sense of the past will be published soon.

[27] P. Geary, *Phantoms of Remembrance: Memory and Oblivion at the End of the First Millennium* (Princeton, 1994), p. 7.

[28] H. Swarzenki, *Monuments of Romanesque Art*, 2nd edn (London, 1974), plate 79.

[29] R. de Lage, 'Les romans antiques et la représentation de la réalité', *Le Moyen Age* 68 (1961),

it was argued a generation ago that at least some of the anachronisms in medieval English Corpus Christi plays were inserted deliberately in order to present 'a critical image of moral and social life as lived in the later Middle Ages'.[30] On what criteria one might decide between these two views it is hard to say. In any case, a sense of distance between antiquity and the present was not openly expressed in the romances any more than it was in the images of the twelfth century.

Three more examples may serve to illustrate the medieval sense of continuity. The first is that of the ruins of Rome, which were regarded as 'marvels', without anyone commenting until the fourteenth century on their date or their architectural style. The second example is that of the Bible, which was interpreted on four levels of meaning, including the literal or historical, but with relatively little interest in what we call the context: who was speaking to whom in what kind of situation. The third example is that of Roman law, which was interpreted like the Bible without reference to the ancient Roman context. One might say that it was adapted to new circumstances, or even that (as has been argued in the case of the romances of Chrétien), that differences between past and present were 'suppressed'. However, this work of adaptation was apparently unconscious. At all events it was not discussed.[31] This point does not imply that there was no awareness at all of change in the law. Medieval lawyers believed that later laws corrected earlier ones, that laws can fall into desuetude and that custom creates rights. The problem is whether these 'ad hoc notions of legal change' should be described as a 'sense of history', and if so, whether it was a sense of history as acute as that of later humanists.[32]

The problem is to find a positive characterization of twelfth-century attitudes. For reasons summarized in the previous section, the phrase 'absence of historical distance' needs to be replaced. The 'sense of historical continuity' is a possible alternative but it is not altogether satisfactory either. A consciousness of continuity is a sophisticated reaction to assertions of discontinuity. In the twelfth century, on the other hand, such assertions had not yet been made.

247–91; J. Frappier, 'La peinture de la vie dans la littérature française du 12e siècle', in *Histoire, mythes et symboles : etudes de littérature française* (Geneva, 1976), pp. 21–54; R. J. Cormier, 'The Problem of Anachronism: Recent Scholarship on the French Medieval Romances of Antiquity', *Philological Quarterly* 53 (1974), 145–57; J. Dunbabin, 'Discovering a Past for the French Aristocracy', in *The Perception of the Past in Twelfth-Century Europe*, ed. P. Magdalino (London, 1993), pp. 1–14 (p. 13); and G. Spiegel, *Romancing the Past: the Rise of Vernacular Prose Historiography in Thirteenth-Century France* (Berkeley, CA, 1993), p. 104.

30 V. A. Kolve, *The Play called Corpus Christi* (Stanford, 1966), p. 104.

31 P. Burke, *The Renaissance Sense of the Past* (London, 1969), pp. 2–6.

32 D. R. Kelley, 'Clio and the Lawyers', in *History, Law, and the Human Sciences: Medieval and Renaissance Perspectives* (London, 1984), pp. 25–49.

In any case, it has been suggested with some plausibility that there are signs of a 'new historic sense' in the twelfth century, at least in scholarly circles. Quoting Guibert of Nogent's denunciation of the corruption of *modernum hoc saeculum,* Etienne Gilson claimed that 'it is in the twelfth century that the West began to become conscious of itself as living in an age distinct from antiquity'.[33] There are a number of indicators of this change in attitude. One is an interest in the 'primitive' Church (*ecclesia primitiva*), often coupled with a sense of the decline of the contemporary Church and a belief in its coming *renovatio*, as in the case of Joachim of Fiore, for example.[34] A second indicator of awareness of change is the debate between *antiqui* and *moderni,* ancients and moderns.[35] Comments on the decline of chivalry parallel those on the decline of the Church and imply a similar division of history into periods.[36]

Ironically enough, the rise of forged documents at this period has also been taken as a sign of historical awareness.[37] The rise of literacy and the increasing need for documentary evidence, even for transactions from earlier periods, encouraged the 'confection' of documents purporting to be centuries old. Literacy also encouraged awareness of the distance between past and present by making more difficult the traditional process of 'structural amnesia', that is, the gradual adaptation of oral traditions about the past to new situations in the present.[38]

The famous concepts of the *translatio imperii* and the *translatio studii* illustrate the twelfth-century sense of the past. They reveal an awareness of change, but also the perception of change as no more than a movement in space, thus preserving the idea of continuity.[39] In the thirteenth century, the 'Roman soldier test' reveals at least one innovative response, that of Nicola Pisano, who (doubtless inspired by Roman sculptures in Pisa) gave his figures a costume rather more Roman than medieval.[40] For literary evidence of a sense of discontinuity or distance, we might turn to the French

[33] E. Gilson, 'Le Moyen Age comme *Saeculum Modernum*', *Lettere Italiane* 24 (1972), 413–20 (p. 415); and Southern, 'Sense of the Past'.

[34] G. B. Ladner, *The Idea of Reform: Its Impact on Christian Thought and Action in the Age of the Fathers* (Cambridge, MA., 1959); G. Miccoli, 'Ecclesiae Primitivae Forma', *Studi Medievali* 1 (1960), 470–98; M. Reeves, *The Influence of Prophecy in the Later Middle Ages: A Study in Joachimism* (Oxford, 1969); G. Olsen, 'The Idea of the *Ecclesia Primitiva* in the Writings of the Twelfth-Century Canonists', *Traditio* 25 (1969), 61–86; *Renaissance and Renewal in the Twelfth Century*, ed. R. L. Benson and G. Constable (Cambridge, MA, 1982).

[35] E. Gössmann, *Antiqui und Moderni im Mittelalter: eine Geschichtliche Standortbestimmung* (Munich, 1974).

[36] R. L. Kilgour, *The Decline of Chivalry* (Cambridge, MA, 1937), p. 4.

[37] H. Fuhrmann, *Einladung ins Mittelalter* (Munich, 1987), pp. 195–210.

[38] Goody, *Literacy*, p. 33; and Clanchy, *Memory to Written Record*, pp. 249–50.

[39] W. Goez, 'Die Anfänge der Historischen Methode-Reflexion in der Italienischen Renaissance und ihre Aufnahme in den Geschichtschreibung des Deutschen Humanismus', *Archiv für Kulturgeschichte* 56 (1974), 25–48.

[40] R. Weiss, *The Renaissance Discovery of Classical Antiquity* (Oxford, 1969), p. 14.

Franciscan Peter John Olivi, whose reflections on the decline of the Church in his day have gained him the title (perhaps exaggerated) of 'historicist'.[41] Again, St Bernard was all too conscious for his own comfort of the decline of the Church in his day.

It is in the fourteenth century, however, that expressions of cultural distance multiply and also extend to the secular world. A sense of renewal in Italian poetry in the time of Guido Cavalcanti and Dante was expressed in the phrase *dolce stil nuovo*. In Paris, two opposing styles of music were described as *ars antiqua* and *ars nova*. In Rome, we find Cola di Rienzo declaring 1347 'year one' of the restored Republic, an affirmation of discontinuity without parallel until the French Revolution.[42] In Spain, Pedro López de Ayala's chronicle has been described as expressing an awareness of the 'gulf separating Spain from its classical past'.[43] In England, a 'classicizing movement' has been detected at this time, expressing itself in forgeries.[44] Most significant of all, Petrarch referred again and again to the cultural distance between the present and his admired antiquity, as well as using his awareness of this distance to show that texts supposed to date from the time of Caesar were later forgeries.[45] It is for these reasons that Petrarch's name appears in the title of this chapter.

The term 'primitive Church' had been in circulation for centuries, as we have seen, but the sense of contrast between a poor and virtuous early Church and a rich and corrupt later Church seems to have been particularly strong at the end of the fourteenth century. For example, we find the Englishman John Wyclif (born *c*.1330) and the Bohemian Jan Hus (born *c*.1369) cutting the seamless web of church history and expressing what has been called a sense of 'estrangement' from the past. Wyclif wrote of the 'decline' of the Church after the riches and power acquired through the *Donation of Constantine*. Hus argued that it was Constantine who had created the papacy itself.[46] A more orthodox churchman, Jean Gerson (born 1363), expressed a similar sense of contrast in the ecclesiastical sphere between the 'primitive Church' and the later Church, as well as a sense of the *varietas temporum*, in other words an awareness that what is appropriate in one epoch is not necessarily so in another.[47] Was it a coincidence that these contemporaries should have found continuity problematic? They may well have been reacting to that blatant breach of continuity, the Avignon papacy, or indeed

41 G. Leff, *Heresy in the Later Middle Ages*, 2 vols. (Manchester, 1967), I, 125.

42 Cola di Rienzo, *Briefwechsel*, ed. K. Burdach and P. Piur, 5 vols. (Berlin, 1912–29), I, 22, 28, 39 and 41.

43 H. Nader, *The Mendoza Family in the Spanish Renaissance* (New Brunswick, 1979), p. 74.

44 B. Smalley, *English Friars and Antiquity* (Oxford, 1940), pp. 230–2.

45 Mommsen, *Medieval*; and Burke, *Renaissance Sense*, pp. 50–4.

46 Leff, *Heresy*, II, 527–8 and 666; P. Meinhold, *Geschichte der Kirchlichen Historiographie* (Freiburg, 1967); and A. Kemp, *The Estrangement of the Past* (New York, 1991), pp. 67–75.

47 L. B. Pascoe, 'Gerson and the Donation of Constantine', *Viator* 5 (1974), 469–85.

to the 'Great Schism' of 1378–1417 (a contemporary term, used by both Froissart and Gerson, for example).

In the secular sphere, the sense of history may also be linked to a sense of decline.[48] The poet Eustache Deschamps, for example, lamented the decline of modern chivalry. Paradoxically, he looked back to the ancient Romans as an example of *bonne chevalerie*, so that he illustrates simultaneously a sense of discontinuity and continuity. Chaucer sometimes committed 'flagrant anachronisms', but on occasion he reveals 'an acute sense of historical change'.[49] A fourteenth-century French monk, Yves de St Denis, using philological methods like Petrarch's, noted anachronisms in the text of 'pseudo-Dionysius', which quoted from texts later than its own supposed date.[50] The names *via antiqua* and *via moderna* were given to opposing trends in philosophy. In similar fashion, the movement associated with the Brethren of the Common Life in the Netherlands was known (by the early fifteenth century if not before) as the *Devotio moderna*.

It was however in fifteenth-century Italy that the sense of anachronism was most acute and most widespread, at least among two groups, humanists and artists, from the Florence of Bruni and Ghiberti to the Rome of Valla or the north Italy of Feliciano and Mantegna. The two groups shared an interest in Pliny's views on ancient art and its growth and decline from youth to old age, knowledge of which encouraged 'a new sense of distance' from the art of the past.[51]

Coluccio Salutati distinguished between the Latin written at the time of Augustus and the Latin written later. Bruni showed his sense of historical context in his translations of Aristotle. The antiquarianism of Flavio Biondo reveals a similar interest in change.[52] So does Poggio Bracciolini's oration in praise of law, in which he contrasts the 'early founders' (*prisci conditores*) with the corruption of the law under the emperors.[53]

Valla's reflections on linguistic anachronism in the text of the *Donation of Constantine* are even more explicit.[54] Valla's remarks might be compared and contrasted with those of Hus (who noted that the term 'pope' did not occur in the Bible, but did not analyse the text of the *Donation*) and also with the common-sense, non-philological rejection on the part of Reginald Pecock.[55]

48 P. Burke, 'European Ideas of Decline and Revival *c.* 1350–1500', *Parerga* 23 (1979), 3–8.

49 M. W. Bloomfield, 'Chaucer's Sense of History', in *Essays and Explorations* (Cambridge, MA, 1970), pp. 13–26.

50 C. Beaune, *Naissance de la nation France* (Paris, 1985), p. 125.

51 S. Settis, 'Did the Ancients have an Antiquity? The Idea of Renaissance in the History of Classical Art', in *Language and Images*, ed. Brown, pp. 27–50 (p. 46).

52 Weiss, *Renaissance Discovery*, pp. 65–73.

53 E. Walser, *Poggius Florentinus: Leben und Werke* (Leipzig, 1914), pp. 258–9.

54 Gaeta, *Lorenzo Valla*.

55 A. B. Ferguson, 'Reginald Pecock and the Renaissance Sense of History', *Studies in the Renaissance* 13 (1966), 147–65; J. M. Levine, 'Reginald Pecock and Lorenzo Valla on the

By this time the Roman soldier test produces even richer results. Ghiberti, for example, gives the soldiers ancient Roman armour.[56] So, more or less, does Piero della Francesca. Mantegna took particular trouble to get the details right, although 'fantastic' elements have been identified in his *Triumphs* as well as exact representations of Roman equipment based on the reliefs on Trajan's Column and other classical sources.[57] Texts confirm the interest in the material culture of the past. The humanist *condottiere* Federigo de Montefeltro once asked the humanist Pope Pius II whether the generals of antiquity wore the same kind of armour as he did ('an prisci duces aeque ac nostri temporis armati fuissent').[58] A sense of anachronism is also revealed in faked objects like the so-called imperial medals which Florentine merchants sold to Jean de Berry, even if they were not particularly accurate imitations.[59] An awareness of differences between past and present was not an Italian monopoly at this time, as the examples of Jan van Eyck (who studied Romanesque architecture and inscriptions) and Richard Rous of Warwick (who studied changes in the design of armour) may remind us.[60]

More difficult to place is the German engraver Israhel van Meckenem, whose image of Judith with the head of Holofernes, made around the year 1500 is presented against a background of contemporary armour and weapons (including cannon). As in the case of the Corpus Christi plays, discussed above, we need to ask ourselves whether this is the naïf anachronism of someone who was not aware of changes in material culture, or the sophisticated anachronism of an artist who wanted to surprise the viewer by introducing topical references to recent events.

DEVELOPMENTS SINCE 1500

In the sixteenth century, the sense of anachronism spread even more widely, enough to undermine the long tradition of exemplarity.[61] In the context of religion, the new sense of the past is revealed by Colet's use of Suetonius to

Donation of Constantine', in *Humanism and History: Origins of Modern English Historiography* (Ithaca, 1987), pp. 54–72.

56 R. Krautheimer and T. Krautheimer-Hess, *Lorenzo Ghiberti* (Princeton, 1956), plates 44 and 50.

57 P. Kristeller, *Andrea Mantegna* (Berlin, 1902), p. 289; and J. M. Greenstein, *Mantegna and Painting as Historical Narrative* (Chicago, 1992), pp. 59–85.

58 A. S. Piccolomini, *Commentarii* (Frankfurt, 1614), p. 190.

59 O. Kurz, *Fakes* (London, 1948); and M. Meiss, *French Painting in the Time of Jean de Berry* (London, 1967).

60 E. Panofsky, *Early Netherlandish Painting*, 2 vols. (Cambridge, MA, 1953), I, 136–9; and J. G. Mann, 'Instances of Antiquarian Feeling in Medieval and Renaissance Art', *Archaeological Journal* 89 (1932), 254–74.

61 Hampton, *Writing*, pp. 54–5, 71 and 75.

interpret St Paul, as it is by Erasmus's insistence on the need to study the history of the time in order to understand the New Testament.

In the context of literature, similar attitudes are revealed in the debate over imitation. Erasmus's *Ciceronianus* employed the argument from anachronism to criticize the writers he called the 'superstitious' imitators of Cicero. The world changes, says one of the speakers in the dialogue, and language changes with it. Since Cicero's time, changes in religion, political organization, laws and customs have made his language an inappropriate model. On the other side, Bembo, for example, was also extremely conscious of changes in language, whether in Latin or in Tuscan.

In the context of the theatre, a similar interest in differences in ways of life and their outward signs is visible. The plays of the Jesuit Bernardino Stefonio, set in ancient Rome, were apparently staged with a careful eye on the accuracy of the costumes.[62] Another Italian playwright, Antonfrancesco Grazzini, went so far as to argue, against the common practice of his day, that Terence and Plautus should not be imitated because 'In Florence we no longer live as they once did in Athens and in Rome: there are no slaves, adopted children are not common'.[63]

In the context of material culture, the report to Leo X by Raphael (or Castiglione or both), in 1519 has been described by a modern art historian as demonstrating 'an unparalleled awareness of architectural history as a sequential development'.[64] Raphael's friend Marco Fabio Calvo made a careful reconstruction of ancient Rome in different periods in a book published in 1527, the year of the sack of the city by imperial troops.[65]

In the case of painting, Vasari praised Giulio Romano for his attention to historical detail. In Ludovico Dolce's dialogue *L'Aretino* (1557), the protagonist, discussing artistic decorum, argues that if the painter wishes to represent one of Caesar's or Alexander's battles, it is inappropriate for him to give the soldiers modern arms and armour ('non conviene che armi i soldati nel modo che si costuma hoggidì').[66] Giovanni Paolo Lomazzo criticized Dürer for architectural anachronism and instructed painters of ancient battles on how to avoid similar mistakes, 'since it is certain that the ancient Romans were dressed differently from moderns' ('andarono vestiti d'altro abito di quello che usano i moderni'). The study of Trajan's Column should set them on the right path.[67]

[62] M. Fumaroli, *L'âge de l'éloquence: rhétorique et 'res literaria' de la renaissance au seuil de l'époque classique* (Geneva, 1980), p. 516.

[63] R. J. Rodini, *Antonfrancesco Grazzini: Poet, Dramatist and Novelliere* (Madison, WI, 1970), p. 42.

[64] J. Onians, *Bearers of Meaning: The Classical Orders in Antiquity, the Middle Ages and the Renaissance* (Cambridge, 1988), p. 258.

[65] Weiss, *Renaissance Discovery*, pp. 95–8; and P. J. Jacks, 'Roman Architecture *all'antica* in 1527', *Art Bulletin* 72 (1990), 453–81.

[66] L. Dolce, *L'Aretino*, ed. M. Roskill (New York, 1968), p. 119.

[67] G. P. Lomazzo, *Trattato dell'arte della pittura*, 2 vols. (Rome, 1844), II, 319 and 556.

By this time a sensitivity to anachronism was no longer the virtual monopoly of Italians. A number of sixteenth-century French scholars demonstrate a sense of historical distance.[68] Indeed, Lomazzo's remarks drew on the work of a French scholar, Guillaume Du Choul, who studied Trajan's Column in order to produce an illustrated book about the military discipline of the Romans.[69] Again, the lawyer-humanists François Baudouin and François Hotman both stressed distance between modern times and ancient Rome, Hotman going so far as to say that the study of Roman law was useless for this very reason.[70]

Linguistic change also attracted attention, just as it had done in Italy. Etienne Pasquier discussed changes in the French language, which has 'assumed different forms according to the diversity of the ages', just like changes in the law.[71] Louis Le Roy made the same point ('Ont les langues comme toutes choses humaines commencement, progrez, corruption, fin') but placed it into the wider context of the 'vicissitudes' of human history.[72] Henri Voisin de La Popelinière drew the methodological moral, criticizing the anachronistic manner in which later historians had represented the early Greeks and Romans, 'non selon le temps et moeurs anciennes; mais selon le siècle de celui qui les recite'.[73]

The new sense of the past both encouraged and was encouraged by the movement for the reform of the Church. The sense of distance between the 'primitive Church' and the modern Church was a stimulus to reform. The more radical the reformer, the earlier he placed the onset of decline. For Luther, it began to be serious in the age of Gregory VII, but the Anabaptists dated the 'fall' of the Church as early as Constantine's establishment of Christianity as the official religion.[74] On the other hand, the rejection of the pope, the destruction of monasteries and other changes consequent on the Reformation led to an increasing sense of estrangement from the past, to which the antiquarian interest in the history of the Church has been seen as a response.[75]

The Catholics too (scholars such as Alfonso and Pedro Chacón, Justus Lipsius, Onofrio Panvinio, and Cesare Baronio) studied Christian antiquities

[68] G. Huppert, *The Idea of Perfect History* (Urbana, IL, 1970), pp. 151–69.

[69] G. Du Choul, *Discours sur la castrametation et discipline militaire des romains* (Lyon, 1557).

[70] F. Baudouin, *De institutione historiae* (Paris, 1561); F. Hotman, *Anti-Tribonian* (Paris, 1630); J. G. A. Pocock, *The Ancient Constitution and the Feudal Law*, 2nd edn (Cambridge, 1987), pp. 11–15 and 22–5; and D. Kelley, *Foundations of Modern Historical Scholarship* (New York, 1970), pp. 116–48 and esp. p. 121.

[71] E. Pasquier, *Les recherches de la France* (Paris, 1596); and Huppert, *Idea*, pp. 28–71.

[72] L. Le Roy, *De la vicissitudes* (Paris, 1575), p. 2; and Huppert, *Idea*, pp. 104–17.

[73] H. La Popeliniére, *L'histoire des histoires* (Paris, 1599); and Huppert, *Idea*, pp. 135–50

[74] J. M. Headley, *Luther's View of Church History* (New Haven, 1963); and Meinhold, *Geschichte*, I, 206 and 211–14.

[75] M. Aston, 'English Ruins and English History', *Journal of the Warburg and Courtauld Institutes* 36 (1973), 231–55; Southern, 'Sense of the Past'; and Kemp, *Estrangement*, pp. 75–97.

in the later sixteenth century, often to reply to the criticisms of the Protestants, but not only for this reason. There was, after all, a tradition of Catholic interest in the so-called 'primitive Church'. The shift in emphasis from pagan Rome to Christian Rome entitles us to speak of a 'Counter-Reformation sense of the past'.[76] Such a phrase may seem strange when one thinks of the effort which Baronius in particular put into demonstrating that the Church was changeless, *Semper eadem*, but the consciousness of continuity in essentials among the scholars just mentioned was very different from the assumption of changelessness which had preceded it. The new concern with changes in the life of the Church (in minor or external matters) was reflected in religious art, with increasingly careful historical reconstructions of the Last Supper and the Crucifixion, inspired by the research of scholars such as Lipsius and Chacón.[77]

As for pagan Rome, a similar interest in appearances and in everyday life is revealed, as we have seen, by studies of the reliefs on Trajan's Column by Du Choul and others. In his case, the discussion of the Dacians, against whom Trajan was campaigning, reveals the gradual awakening of interest in the appearance and customs of the 'barbarians' against whom the Romans fought, an interest which may well have been stimulated by accounts of the indigenous inhabitants of America. Early seventeenth century representations of ancient Britons and Picts, for example, are indebted to John White's paintings of the Indians of Virginia.[78]

In the Netherlands, Spain, and elsewhere, attempts were made to present Greeks and Romans and even characters from the Middle Ages in clothes which were not contemporary. When the Emperor Charles V first visited Spain, he was met by knights of the Golden Fleece dressed, according to a contemporary chronicler, Laurencio Vital, 'according to the fashion of past times' (*según la moda de tiempos passados*).[79] When Prince Philip (later Philip II) entered Antwerp in 1549, he was met by three other Philips from different epochs, each dressed, apparently, in the clothes of the period: Philip of Macedon in 'Greek clothes' (*grieksche cleedinge*), St Philip in 'apostolic clothes' (*apostelsche cledinge*), whatever those were supposed to have been, and Philip of Burgundy in 'his old style of clothes' (*sijne oude maniere van cleedingen*).[80] The dramatist Lope de Vega, in his *New Art of Writing Plays*, mocked the 'barbarisms' of his day, such as showing ancient Romans in trousers. The stage directions to Cervantes' *Siege of Numancia*

[76] J. Orella, *Respuestas a las centurias* (1976); and E. Cochrane, *History and Historians in the Italian Renaissance* (Chicago, 1981), especially ch. 16.

[77] A. Blunt, 'The Triclinium in Religious Art', *Journal of the Warburg Institute* 2 (1938–9), 271–6.

[78] T. Kendrick, *British Antiquity* (London, 1950), pp. 121–2.

[79] Quoted in F. Checa, *Carlos V y la Imagen del héroe en el renacimiento* (Madrid, 1987), p. 191.

[80] H. van de Waal, *Drie eeuwen Vaderlandsche Geschied-Uitbeelding 1500–1800* (The Hague, 1952), p. 26.

describe Scipio's soldiers as 'armed in the ancient manner, without arquebuses'.[81]

Possible differences between renaissance scholars and modern ones come into focus if we examine the question of motive. Recent work on renaissance forgeries and detections of forgeries, from Annius of Viterbo to Isaac Casaubon, has emphasized 'the distance which lies between the humanists' exposures of fakes and those of the modern philologist'. For example, the new methods were sometimes used to create forgeries as well as to expose the forgeries of others.[82] The parallel between the faking of ancient works of art and the forging of classical texts (in some cases as a demonstration of skill, in others for profit) appears to be a close one.[83] Ficino's enthusiasm for the Hermetic writings, for example, seems to have blinded him to the anachronisms which made them seem obvious forgeries to Casaubon. The humanists were extremely conscious of the difference between past and present. Until the later sixteenth century, they were much less aware of development, especially between early and late antiquity.[84]

THE EARLY SEVENTEENTH CENTURY

At this point, in the early seventeenth century, a discussion of the 'renaissance sense of the past' may reasonably come to an end, though without any implication that changes in the perception of the past ceased to take place. Recent research has shown both the importance and the development of an increasingly acute sense of historical distance after 1650.

However, the interest in change and in anachronism in the early seventeenth century deserves more emphasis than it has traditionally received. For example, if some Protestants of the time claimed that the example of the Old Testament should be imitated in almost every respect, others rejected the idea as impossible. James I declared that the law of Moses was only the law of Israel, 'only fit for that country, that people and that time'.[85] In politics, Francisco de Melo criticized Olivares for his imitation of the ancients, 'as if Tacitus, Seneca . . . of whom he took counsel, would not have altered their

[81] N. D. Shergold, *A History of the Spanish Stage from Medieval Times until the End of the Seventeenth Century* (Oxford, 1967), p. 190; and S. M. Newton, *Renaissance Theatre Costume and the Sense of the Historic Past* (London, 1975).

[82] Grafton, *Defenders*, p. 79; and Goez, *Anfänge.*

[83] O. Kurz, *Fakes* and J. Alsop, *The Rare Art Traditions* (New York, 1982).

[84] Contrast Z. S. Schiffman, 'Renaissance Historicism Reconsidered', *History and Theory* 24 (1985), 170–82, and A. B. Ferguson, *Clio Unbound: Perception of the Social and Cultural Past in Renaissance England* (Durham, NC, 1979), esp. ch. 10.

[85] K. Thomas, 'The Puritans and Adultery', in *Puritans and Revolutionaries*, ed. D. Pennington and K. Thomas (Oxford, 1978), pp. 257–82 (p. 269).

views if they had been alive now'.[86] In a similar fashion to La Popelinière, Sir Henry Spelman's study of parliaments criticized the projection of the present on to the past: 'succeeding ages viewing what is past by the present conceive the former to have been like to that they live in.'

In literature and art as in politics, expressions of an awareness of differences between historical epochs are easy to find. The critic Jean Chapelain made a study of medieval romances in which he noted that 'Notre manière de plaire aux dames et de leur persuader que nous les aimons est tout contraire à celle des vieux âges.'[87] In the case of painting, Salvator Rosa attacked the anachronisms of other artists, while Rubens was praised by the scholar Claude Peiresc for the care with which he rendered classical costume in his *Allocution of Constantine*. Again, Nicholas Poussin (who made a careful study of Trajan's Column and the works of Du Choul and Chacón), showed the apostles reclining at the Last Supper on the model of the ancient Roman *triclinium* and the Jewish Seder. Like Mantegna and Raphael, Poussin was a friend of scholars, of Cassiano del Pozzo, for instance, and of André Félibien, the French historian of painting who lived in Rome from 1647 to 1649. He was one of the first to recognize differences between Greek and Roman art, instead of treating classical antiquity as undifferentiated. It is for these reasons, and to emphasize the point that the discovery of the 'otherness' of the past was the joint work of scholars and artists, that Poussin's name appears in the title of this paper.[88]

It was, finally, around 1650 that the term 'anachronism' (*anachronismus, anacronismo, anachronisme*) began to come into use in Latin, Italian, French and English, employed by such scholars as the Jesuits Denys Petau and Sforza Pallavicino, and by the bishops Jacques Bossuet and Pierre-Daniel Huet. An anachronism was a mistake made in the course of 'synchronism', in other words the attempt to translate from one chronological system into another. To speak of the sense of anachronism of Mantegna or Erasmus is therefore, literally speaking, anachronistic.

All the same, the rise of what it remains convenient to call a sense of anachronism among a group of scholars and artists, especially Italians, in the fifteenth and sixteenth centuries was an important event in European cultural history. As we have seen, the shift which took place in this period was part of a longer trend which goes back at least as far as the twelfth century as well as forward to the present. However, these qualifications to the thesis put forward by Warburg, Panofsky and others early this century have refined it but they have not destroyed it.

[86] J. H. Elliott, 'The Statecraft of Olivares', in *The Diversity of History*, ed. J. H. Elliott and H. G. Koenigsberger (London, 1970), pp. 117–48 (p. 124).

[87] J. Chapelain, *Opuscules critiques* (Paris, 1936), p. 237.

[88] E. Cropper and C. Dempsey, *Nicolas Poussin: Friendship and the Love of Painting* (Princeton, 1996).

INDEX

YORK MEDIEVAL PRESS: PUBLICATIONS

God's Words, Women's Voices: The Discernment of Spirits in the Writing of Late-Medieval Women Visionaries, Rosalynn Voaden (1999)

Pilgrimage Explored, ed. J. Stopford (1999)

Piety, Fraternity and Power: Religious Gilds in Late Medieval Yorkshire 1389–1547, David J. F. Crouch (2000)

Courts and Regions in Medieval Europe, ed. Sarah Rees Jones, Richard Marks and A. J. Minnis (2000)

Treasure in the Medieval West, ed. Elizabeth M. Tyler (2000)

Nunneries, Learning and Spirituality in Late Medieval English Society: The Dominican Priory of Dartford, Paul Lee (2000)

Prophecy and Public Affairs in Later Medieval England, Lesley A. Coote (2000)

The Problem of Labour in Fourteenth-Century England, ed. James Bothwell, P. J. P. Goldberg and W. M. Ormrod (2000)

New Directions in Later Medieval Manuscript Studies: Essays from the 1998 Harvard Conference, ed. Derek Pearsall (2000)

Cistercians, Heresy and Crusade in Occitania, 1145–1229: Preaching in the Lord's Vineyard, Beverly Mayne Kienzle (2001)

Guilds and the Parish Community in Late Medieval East Anglia, c.1470–1550, Ken Farnhill (2001)

The Age of Edward III, ed. J. S. Bothwell (2001)

York Studies in Medieval Theology

I *Medieval Theology and the Natural Body*, ed. Peter Biller and A. J. Minnis (1997)

II *Handling Sin: Confession in the Middle Ages*, ed. Peter Biller and A. J. Minnis (1998)

III *Religion and Medicine in the Middle Ages* ed. Peter Biller and Joseph Ziegler (2001)

York Manuscripts Conferences

Manuscripts and Readers in Fifteenth-Century England: The Literary Implications of Manuscript Study, ed. Derek Pearsall (1983) [Proceedings of the 1981 York Manuscripts Conference]

Manuscripts and Texts: Editorial Problems in Later Middle English Literature, ed. Derek Pearsall (1987) [Proceedings of the 1985 York Manuscripts Conference]

Latin and Vernacular: Studies in Late-Medieval Texts and Manuscripts, ed. A. J. Minnis (1989) [Proceedings of the 1987 York Manuscripts Conference]

Regionalism in Late-Medieval Manuscripts and Texts: Essays Celebrating the Publication of 'A Linguistic Atlas of Late Mediaeval English', ed. Felicity Riddy (1991) [Proceedings of the 1989 York Manuscripts Conference]

Late-Medieval Religious Texts and their Transmission: Essays in Honour of A. I. Doyle, ed. A. J. Minnis (1994) [Proceedings of the 1991 York Manuscripts Conference]

Prestige, Authority and Power in Late Medieval Manuscripts and Texts, ed. Felicity Riddy (2000) [Proceedings of the 1994 York Manuscript Conference]

Middle English Poetry: Texts and Traditions: Essays in Honour of Derek Pearsall, ed. A. J. Minnis (2001) [Proceedings of the 1996 York Manuscript Conference]

www.ingramcontent.com/pod-product-compliance
Ingram Content Group UK Ltd.
Pitfield, Milton Keynes, MK11 3LW, UK
UKHW020843290926
15470UKWH00004B/1523

9 781903 153086